The Southdown Sheep Flock Book

Volume 1

by Southdown Sheep Breeders Association

with an introduction by Jackson Chambers

Self Reliance Books

Get more historic titles on animal and stock breeding, gardening and old fashioned skills by visiting us at:

Introduction

I am pleased to present yet another practical title on breeding and raising livestock.

The work is in the Public Domain and is re-printed here in accordance with Federal Laws.

As with all reprinted books of this age that are intended to perfectly reproduce the original edition, considerable pains and effort had to be undertaken to correct fading and sometimes outright damage to existing proofs of this title. At times, this task is quite monumental, requiring an almost total "rebuilding" of some pages from digital proofs of multiple copies. Despite this, imperfections still sometimes exist in the final proof and may detract from the visual appearance of the text.

I hope you enjoy reading this book as much as I enjoyed making it available to readers again.

Jackson Chambers

The Celebrated Blackfaced Ram "CAIRNTABLE,"
Bred by CHARLES HOWATSON, of Durnel, and sold for **SIXTY POUNDS**,
To J. N. FLEMING, of Keil   July, 1870

TABLE OF CONTENTS.

Southdown Sheep Breeders' Association.

PRESIDENT.
Mr. EDWIN ELLIS.

VICE-PRESIDENTS.
His Grace the Duke of Northumberland.
The Right Hon. Earl Bathurst.
Sir Thomas Barrett-Lennard, Bart.
Mr. Laurence Cave.
Mr. H. F. Locke-King.
Mr. A. de Murietta.
Mr. W. Trotter.

MEMBERS OF COUNCIL.

Mr. Robt. Anderson, Junr.	Mr. H. F. Locke-King.
Sir Thos. Barrett-Lennard, Bart.	Mr. F. N. Hobgen.
Mr. J. W. Baxendale.	Mr. H. Padwick.
Mr. T. Botting.	Mr. H. Penfold.
Mr. W. W. Chapman.	Mr. E. M. Synge.
Mr. Newton Clayton.	Mr. W. Toop.
Mr. C. Czarnikow.	Mr. W. Trotter.
Mr. W. Flux.	Mr. D. Waddington.
Mr. H. Jay.	Mr. P. H. Weber.

INSPECTION COMMITTEE.

Sir T. Barrett-Lennard, Bart.	Mr. N. Clayton.
Mr. R. Anderson.	Mr. F. N. Hobgen.
Mr. T. Botting.	Mr. H. Penfold.
Mr. W. W. Chapman.	Mr. W. Toop.

FINANCE COMMITTEE.

Mr. Edwin Ellis.	Mr. E. M. Synge.
Mr. H. F. Locke-King.	Mr. W. Trotter.

EDITING COMMITTEE.

Mr. N. Clayton.	Mr. W. W. Chapman.

AUDITORS.

Mr. H. Penfold.	Mr. W. Toop.

HON. SOLICITORS.
Messrs. Flux, Son & Co., 3, East India Avenue, London, E.C.

HON. TREASURER.
Mr. H. F. Locke-King.

BANKERS.
Capital and Counties Banking Co., 39, Threadneedle Street, London, E.C.

SECRETARIES.
Mr. Newton Clayton, *Hon. Secretary.*
Mr. W. W. Chapman, *Secretary.*

ASSOCIATION'S OFFICES.
27, Baker Street, Portman Square, London, W.

LIST OF MEMBERS.

ANDERSON, ROBERT, Junr., The Bartons, Cirencester.

BARRETT-LENNARD, Sir THOMAS, Bart., Woodingdean, Brighton.

BANNISTER, Thomas, Hayward's Heath, Sussex.

BATHURST, Rt. Hon. Earl, Cirencester.

BAXENDALE, J. W., Hursley Park, Winchester.

BLYTH, JAMES, Blythwood, Stanstead, Essex.

BOTTING, THOMAS, Shalford, Guildford.

CAVE, LAURENCE, Ditchat Park, Petersfield.

CHAPMAN, W. W., Wadhurst Park, Sussex.

CLAYTON, NEWTON, Selsey, Chichester.

CZARNIKOW, C., Effingham Hill, Dorking.

ELLIS, EDWIN, Summersbury, Shalford, Guildford.

FLUX, W., 3, East India Avenue, London, E.C.

HABIN, H., Birdham, Chichester.

HOBGEN, F. N., Appledram, Chichester.

JAY, H., Fittleworth, Pulborough.

LOCKE-KING, H. F., Brooklands, Weybridge.

MURIETTA, A. DE, Wadhurst Park, Sussex.

NORTHUMBERLAND, His Grace the Duke of, Albury, Guildford.

PAGE-WOOD, CHARLES, Wakes Hall, Wakes Colne, Essex.

PADWICK, H., Manor House, West Thorney, Emsworth.

PAGHAM HARBOUR COMPANY, Selsey, Chichester.

PENFOLD, HUGH, Selsey, Chichester.

PERKINS, Mrs., Saham Hall, Walton, Norfolk.

SIMMONS, H. M., St. Leonard's Villa, Eastbourne.

SYNGE, E. M., Eastlands, Weybridge.

TOOP, W., Aldingbourne, Chichester.

TROTTER, W., Sydenhurst, Chiddingfold.

WADDINGTON, D., Manor House, Lavant, Chichester.

WEBER, F. H., Grantham, Chiddingfold.

WOODS, WALTER, Lavant, Chichester.

Extracts from Memorandum and Articles of Association.

1. The name of the Association is "THE SOUTHDOWN SHEEP BREEDERS' ASSOCIATION."

2. The Registered Office of the Association will be situate in England.

3. The objects for which the Association is established are:—

(a) The encouragement of the breeding of Southdown sheep at home and abroad, and the maintenance of the purity of the breed.

(b) The establishment and publication of a flock book of recognised and pure-bred sires which have been used, and of ewes which have been bred from, and of such other flock books (if any) which the Council may think fit, and the annual registration of the pedigrees of such sheep as are proved to the satisfaction of the Council to be eligible for entry.

(c) The periodical compilation and publication of a statement of transactions connected with the breed, such as particulars relating to shows, sales and other transactions, with such other general information with reference to the breeding and management of sheep, and to sheep farming, as the Council may think fit.

(d) The arrangement of classes, and the donation or augmentation of prizes and awards of certificates of merit at various shows, and the appointment or recommendation of judges thereat.

(e) The investigation of cases of doubtful and suspected pedigrees.

(f) The undertaking of the arbitration upon and settlement of disputes and questions relating to or connected with Southdown sheep and the breeding thereof, and for other subsidiary purposes.

(g) To provide rooms and other facilities for the holding and conducting of Meetings for the objects or work of the Association.

(h) To purchase, take on lease, hire, receive by way of gift, or otherwise acquire, and also to sell, let, or dispose of, any real or personal property for the purposes of the Association, subject to the provisions of the 21st section of " The Companies Act, 1862."

(i) To borrow from time to time any moneys required for the purposes of the Association upon such security as may be determined, but so that the moneys at any one time owing shall not exceed £100 beyond what the Council shall estimate to be the value of any real or personal property of the Association at the time of borrowing.

(j) To promote information with reference to sheep breeding by lectures, discussions, books, correspondence, or otherwise, and for any of the objects of this clause—of paragraph 3—to co-operate with any university or college, or professor or lecturer thereof.

(k) To consider all questions affecting the interest of sheep breeders, and to initiate and watch over, and, if necessary, to petition Parliament or any local authority, or promote deputations in relation to general measures affecting sheep breeders, and to procure changes of law affecting sheep, and the promotion of improvements in the administration of the law affecting sheep.

(*l*) The doing of all such other lawful things as are incidental or conducive to the attainment of the above objects or any of them.

Any person desirous of becoming a Member of the Association shall be proposed by one Member of the Association and seconded by another Member of the Association, and elected by a majority of the Members present at a Council or General Meeting of the Association. Any Member may at any time by notice in writing to the Secretary resign his Membership, but such resignation shall not free him from liability to pay any annual subscription or other money which may be due from him to the Association and unpaid.

The rights and privileges of every Member of the Association shall be personal to himself, and shall not be transferable or transmissible.

Every Member of the Association shall, provided he has paid his annual subscription for the current year, or has compounded for the same, as hereinafter mentioned, be entitled to a copy of each of the publications of the Association without further charge.

OBLIGATIONS OF MEMBERS.

Every Member on joining the Association shall pay to the funds of the Association an entrance fee, and shall also pay an annual subscription, or may, at his option, either on his joining the Association or at any other time afterwards, pay a lump sum by way of composition. In lieu of such annual subscription, such entrance fee, subscription, and composition to be according to a scale to be from time to time determined by the Council. The annual subscription shall be payable in advance on the First day of January in each year, unless the Member who would otherwise have been liable to pay the same shall give three months' notice in writing to the Secretary, before that date, of his intention to withdraw from the Association.

Every Member shall observe all lawful bye-laws, regulations, and orders of the Council, for the government and work of the Association, and for Shows held by or in connection therewith, and pay all fines and forfeits which the Council shall in pursuance of their powers impose.

The management of the business of the Association shall be vested in the Council, who, in addition to the powers and authorities by statute or by these Articles expressly conferred upon them, may exercise all such powers and do all such acts and things as are, or shall be, by statute or these Articles, directed or authorised to be done by the Association, and not hereby expressly directed to be done by the Association in General Meeting assembled, but subject nevertheless to the statutory provisions and these Articles, and subject also to such (if any) regulations as may be from time to time determined by any Special Meeting of the Association, but no such regulation shall invalidate any prior act of the Council which would have been valid if the regulation had not been made.

BYE-LAWS.

SUBSCRIPTION OF MEMBERS.

For the present the following shall be the respective amounts due from Members under the terms of the Ninth Article of Association :—

	£	s.	d.
Entrance Fee	1	1	0
Annual Subscription	1	1	0
Life Composition in lieu of Annual Subscription...	15	15	0

In case of a firm, one Subscription will be deemed sufficient to entitle the partners therein to the joint rights and privileges of one individual membership, and either partner shall be entitled (subject to Article 26 of the Articles of Association) on behalf of the firm, to exercise such rights and enjoy such privileges.

COUNCIL.

Meetings of Council shall be held each year at the Offices of the Association during the show of the Smithfield Club in London, and at such other times and places as the Council shall from time to time appoint.

The Council, at its first meeting after the Annual General Meeting, shall appoint from among its Members an Editing Committee, a Finance Committee, and a Committee of Inspection. The Flock Book shall be published annually by the first day of July, and shall contain a register of Southdown Sheep, a statement of the results of the principal shows and sales during the year, and any other information which, in the opinion of the Council, shall be of interest to breeders. The Editing Committee shall be responsible to the Council for the publication of the Flock Book. No sheep shall be eligible for entry in the Flock Book, except such as shall be proved to the satisfaction of the Council to be pure-bred Southdowns.

The price of the Flock Book to non-Members shall not exceed 10s. 6d.

The charge for entering animals in the Flock Book shall be:—

	£	s.	d.
To Members, For each individual Ram or Ewe	0	2	6
„ For Ewes *en bloc*, the flock	2	2	0

FINANCE COMMITTEE.

All payments on behalf of the Association shall be made by drafts on the Treasurers, signed by the Chairman of the Finance Committee (or by two members thereof) and countersigned by the Secretary.

STANDING COMMITTEE.

Meetings of Committees shall be summoned at such times and places as the Chairman of each Committee shall appoint.

THE SOUTHDOWN SHEEP.

By Newton Clayton and W. W. Chapman.

"Small in size, but great in value," is still characteristic of the true Southdown, and it would be a sad event in the history of this oldest and purest of breeds if size should become preferred to quality, and mere weight of mutton to character and good wool.

The antiquity of the Southdown is unquestioned, and there are to-day flocks of Southdown sheep cropping the short pasture of the Sussex downs, whose traditions are lost in the centuries. The fame of its mutton would appear to be almost as ancient; we read that the "Merry Monarch" "loved Southdown mutton," and in 1798 the Emperor of Russia bought, at a cost of 300 guineas, two rams from Mr. Ellman, of Glynde, to improve the flavour and quality of the northern mutton.

Originally the Southdown was light in the forequarters, narrow chested, thin necked, and rather long in the leg; but it was hardy, and would thrive on pasture that would starve many other breeds of sheep.

Messrs. Ellman, of Glynde, and at a later period Mr. Jonas Webb, seem to have done for the Southdown what Bakewell did for the Leicester, and by judicious selection to have produced a sheep symmetrical in form, of hardy constitution, with good wool and character, and the best of mutton.

In a good Southdown we look for a head wide and level between the ears, with no sign of slug or dark poll; eyes large, bright and prominent; ears of medium size, covered with short wool; face full, not too long from eyes to nose, and of one even mouse colour, not approaching black nor speckled with white, under jaw, light; neck wide at base, strong and good; shoulders, well put in, the top level with the back; chest, wide and deep, "thick through the heart"; fore flanks fully developed; ribs wide sprung and "well·ribbed up"; back level, with wide and flat loin, the whole covered with firm flesh; flanks, deep and full; rump, wide, long and well turned; tail, large, and set on almost level with the chine; thighs, full, well let down, with deep wide twist, ensuring a good leg of mutton; legs, a mouse colour, and "outside the body," the whole of which should be covered with a fine, close and even fleece down to hocks and knees, and right up to the cheeks, with full foretop, but there should be no wool under the eyes or across the bridge of the nose. The skin should be of a delicate and bright pink, the carriage gentlemanly, and the walk that of a thoroughbred.

A breed of sheep that will put on a pound of mutton at a less cost and of greater value than any other, must be the most profitable; therefore, the Southdown is the real rent-paying sheep. It will keep in good condition on bare pasture, and on generous diet will mature as quickly as any breed.

Its value for "crossing" is acknowledged. It occupies the same position amongst sheep as the Shorthorn does in cattle, and it has been freely laid under contribution to produce the "improved" Hampshire and Oxford Downs, Shropshires, &c.

The many excellencies of the Southdown—the hardy constitution, the adaptability to almost any climate, the habit of thriving on bare pasture, the generous return for good feeding, the comparative immunity from foot rot and less liability to "fly" (from the density of its fleece), the general aptitude to improve other breeds by "crossing," the beauty of character, the fine quality of mutton and the excellence of its wool, only require to become known to be thoroughly appreciated by home and foreign flock owners.

SOUTHDOWN FLOCKS.

N.B.—With the exception of Flocks 1 and 2, which are sold, every sheep in Flock Book is ear-marked with the Association's stamp, followed by its Flock number.

FLOCK No. 1.

Messrs. DE MURIETTA, Wadhurst Park, Wadhurst, Sussex.

Southdowns have been kept at Wadhurst Park from the time the estate was purchased, about 25 years ago; but in 1887 it was determined to make this, if possible, one of the best flocks of the kingdom for purity of blood, character, and wool, and, in order to be prepared for the future registration of the Southdown Sheep, it was decided to keep a careful register of each and every sheep in the flock. In this year (1887) 125 Ewes were purchased at C. J. Carew Gibson's Sandgate sale, and 15 Ewes at Mr. Gorringe's annual sale, Kingston-on-Sea.

In 1888 100 Ewes were purchased from the famous flock of Col. Sir Nigel Kingscote, and at the dispersal sale of Mr. H. Penfold's flock at Selsey 31 of the best of the Ewes were obtained. In 1889 15 Ewes came from the sale of H.R.H. the Prince of Wales, and at Mr. H. Webb's first sale 65 of the best of the Ewes were purchased. At his final sale in 1890, 20 Ewes were obtained, and in this and the following year, 1891, Ewes were purchased from the flocks of Sir William Throckmorton, Mr. F. Barchard, and Mr. Stewart Hodgson.

The Sires used since 1887 are as follows:—

In 1887 two Rams from Mr. C. G. Carew Gibson, and one hired from Mr. Hugh Gorringe for the season.

In 1888 two Rams hired of Mr. Henry Webb, No. 24, purchased at Mr. Penfold's sale, and No. 7 hired from Goodwood.

In 1889 Mr. Penfold's No. 24; Mr. Webb's No. 39 of 1889, "Gloucester" (sire of "Cambridgeshire"); No. 7 of 1888; and a Ram Lamb, No. 10, from Mr. Heasman's sale at Arundel.

In 1890 Mr. Penfold's No. 24, Mr. Webb's No. 39 of 1889, and also No. 44 of 1890; Mr. Heasman's Ram Lamb No. 10, a "Throckmorton" Ram, and several home-bred ones.

In 1891 Mr. Penfold's No. 24, Mr. Webb's No. 39 of 1889, Sir N. W. Throckmorton's No. 1 of 1891 sale, two of Ellis' Rams, several home-bred Rams; and 20 Ewes were specially selected and sent for service to Mr. Ellis' celebrated Ram, "Baronet."

At the dispersion of the flock (every sheep being sold) in 1892—a year in which Southdowns generally realised very low prices—the averages were:—

47 Rams	...	...	...	£10	8	0 per head.
80 Ram Lambs	...	...		4	7	0 ,,
362 Ewes	..	...	over	3	0	0 ,,

Ewes	... ...	269		
Shearling Ewes	... ...	99	Rams and Ram Lambs	136

N.B.—This flock was sold on August 25th, 1892.

FLOCK No. 2.

W. TROTTER, Esq., Sydenhurst, Chiddingford, Surrey.

The Sydenhurst Flock was founded several years ago, with a number of carefully selected Ewes, among which were twenty choice Shearlings and a Ram from Sir N. W. Throckmorton's old-established flock at Buckland. Purchases of Ewes were also made from the Earl of Bathurst at Cirencester, from Mr. Edwin

Ellis's prize-winning flock, and Mr. Hugh Penfold of Selsey, Chichester; the Buckland Ram was followed by Sheep from Mr. J. Stewart Hodgson, Mr. W. Toop of Aldingbourne, and Mr. Henry Webb.

Ewes 158
Shearling Ewes 65   Rams and Ram Lambs   42

N.B.—This flock was sold on August 3rd, 1892.

FLOCK No. 3.

PAGHAM HARBOUR CO., Selsey, Chichester. Address: Newton Clayton, Selsey, Chichester.

This flock was established by the purchase, in 1879, of 70 draft Ewes from the old-established flock of Mr. Hugh Penfold (No. 4 in Flock Book). In 1880 160 Ewes were bought at the sale of Thomas Saxby, who had bred largely from Messrs. Ellman. On three subsequent occasions Ewes were obtained from Penfold—three lots of draft Ewes, 120 in all, were added from the old Goodwood flock, and selected Ewes have also been procured from the flocks of Carew Gibson, Heasman, Hobgen, &c.

Recognizing the excellent character and wool of the old Selsey flock, Penfold's best Rams have been chiefly used every year, all his choicest strains having been brought into requisition, but for several years rams have also been used bred by Mr. Henry Webb, also three sons of "General Favourite," which were worked three years.

The farms being situate on a promontory jutting into the English Channel, without shelter, particular attention has always been given—not only to true Southdown character but also to hardiness of constitution and density of fleece. Many prizes have been taken for Sheep at the "Royal" and other shows, but wool has only been once exhibited, gaining 1st Prize at the "Jubilee Royal" show, Windsor, 1889.

Ewes 1070   Rams 40

FLOCK No. 4.

HUGH PENFOLD, Selsey, Chichester.

This flock has been established over a century, and has ever been celebrated for its Southdown character and for the excellence of its wool. The Sheep have for the past 30 years been very successful in the showyard.

Since the foundation of the flock Rams have been selected with the greatest care, and with special regard to purity, character, and wool, from the celebrated flocks of Messrs. Ellman, Henry Webb, Duke of Richmond, H. P. Hart, John Walters, Lord Walsingham, Hugh Gorringe, and other breeders.

Ewes 250
Shearling Ewes 80   Rams and Ram Lambs   32

FLOCK No. 5.

HARRY HABIN, Birdham, Chichester.

Old Ewes... 227
Shearling Ewes 234   Rams 6

FLOCK No. 6.

FRANCIS NEALE HOBGEN, Manor Farm, Appledram, Chichester.

This flock was started in 1833 by a draft of 200 Ewes from the flock of the Duke of Richmond, and Rams bred by the late William Oliver, John Pinnix, and

Hugh II. Penfold were used, the flock being then the property of Mr. Charles Hobgen, the present owner's uncle, who gave up business 13 years ago, when the entire flock came into the hands of the present owner, who first introduced blood from the flocks of Messrs. John Humphry and Carew Gibson, and since then from other pure bred flocks. In 1889 Rams were used of Messrs. Carew Gibson's blood, Edwin Ellis, Hugh Gorringe, and Hugh Penfold. In 1890 Rams were used from the flocks of the Pagham Harbour Company, Hugh Penfold, and E. Ellis, and in 1891 Rams used were from the flocks of E. Hobgen, Lord Alington, Mr. Lucas, Mr. A. T. Newman, and Mr. W. A. Hammond.

Ewes	...	...	... 350	Rams	21
Shearling Ewes ...	... 114				

FLOCK No. 7.

SIR THOMAS BARRETT-LENNARD, Bart., Woodingdean, Brighton.

This flock was founded in 1890 by Ewes bought of Mr. Brown, and Ewes have been since purchased from Messrs. Ellis, Lucas, and Henry Webb, and Rams from Messrs. Henry Webb, Gorringe, Ellis, Lucas, and Colman.

Ewes	...	...	... 205	Rams	10
Shearling Ewes ...	... 105				

FLOCK No. 8.

SIR THOMAS BARRETT-LENNARD, Bart., Belhus, Romford, Essex.

This flock was founded in 1857 by the present owner's grandfather, who had used Webb's Rams for some years, and the present owner introduced into the flock from time to time Ewes from the flocks of Mr. Rigden, Sir W. Throckmorton, Earl of Bathurst, and others. At the late Mr. Jonas Webb's great sale as much as 35 guineas per head was given for 5 Ewes, and lately Ewes have been purchased from the flocks of Messrs. Ellis and Lucas.

Rams from the flocks of Messrs. Rigden, Hart, Jonas Webb (at whose sale 140 guineas was paid for a Ram), Ellis, and Henry Webb.

In 1865 the Gold Medal presented by the late Emperor of France at Poissy, near Paris, was won by Sheep from this flock, and numerous prizes have been won at Essex and Suffolk County shows.

Ewes ...	..	...	...	44
Shearling Ewes	...	...		6

FLOCK No. 9.

WILLIAM TOOP, Aldingbourne, Sussex.

The flock was established in 1883 from Ewes bred by Messrs. Ford, Elliott, and Padwick, and has been replenished by Ewes from the flocks of Hugh Penfold, Henry Humphrey, Carew Gibson, William Woodman, Henry Upton, and Henry Webb.

Rams have been used bred by Hugh Penfold, Botting, Newman, and the Duke of Richmond, and numerous prizes have been won at the principal shows—including Champion and Gold Medal at Smithfield.

Ewes	...	...	... 185	Rams	42
Shearling Ewes ...	... 90				

FLOCK No. 10.

H. F. LOCKE-KING, Oakhurst Manor, Billingshurst.

Though Southdowns had been kept for many years on Mr. Locke-King's Sussex farms, the establishment of the present flock may be said to date from the year 1887, when selections of Ewes were made from the flocks of Messrs. Gibson, Stanford, Shelley, and Drewitt. In the following year, 1888, a Ram, No. 36, at Mr. Penfold's sale was purchased for use in the flock and did good service, many of the Ewes now in the flock being by him. At the same sale two pens of Ewes were obtained, and during the summer additions made from the flocks of Messrs. Ellis, H. Gorringe, Drewitt, Emery, and Sir A. Hornsby. Besides the Ram from Mr. Penfold, two Rams from Mr. Ellis were used this season.

In 1889, two Rams were obtained from the well-known flock of J. J. Colman, M.P., one of them a son of Old Norwich, both Sheep being of excellent quality and with first-class wool. During the same summer purchases were made at the dispersion of Lord Onslow's flock and at the sales of Messrs. Ellis, Heasman, and Upton.

In 1890, a few Ewes were purchased from Mr. Gibson, and in 1891, a pen of Ewes was added from Mr. Lucas' flock and a son of Earl obtained at Mr. Ellis' sale.

Ewes 300	Rams 9
Shearling Ewes 101	

FLOCK No. 11.

H. F. LOCKE-KING, Brooklands, Weybridge.

This flock, in 1891, consisted entirely of Ewes purchased from the Marquis of Bristol, at the Ickworth sale, in July, 1891. There are records of Southdowns at Ickworth as far back as 1807, and throughout its existence care was taken to obtain none but the best and purest blood. For some ten years after 1849, Sheep were purchased from Mr. John Ellman, of Glynde, and afterwards from the flocks of Mr. Rigden and Lord Walsingham. Sheep were also obtained from the flocks of Mr. Humphrey, of Ashington, and of Mr. Penfold, and Rams purchased at the Babraham sales in 1861 and 1862. For the Ickworth flock, from 1872, Rams were hired by Lord Bristol from Mr. Henry Webb, of Streetly Hall, and Rams and Ewes were also purchased at his sales.

One of these Rams, No. 27 in the 1889 sale, was bought at the Ickworth sale, and was used in the Weybridge flock in the past season. The other sires for the season of 1891 were a son of Mr. Webb's, No. 24 in 1889, purchased at Mr. Stewart Hodgson's sale, and a son of Favourite, purchased at the Chichester Ram sale from the flock of Mr. Heasman.

Ewes 171	Rams 4
Shearling Ewes 38	

FLOCK No. 12.

EDWIN ELLIS, Summersbury, Shalford, Guildford.

The Summersbury Stock was established in 1879 by the purchase of carefully selected Ewes from the flock of Messrs. Botting, where the old Rigden blood had been freely used, and which had also been replenished by some of Colman's best Rams.

In 1885, at the dispersal of Lord Walsingham's flock, the Rams Merton and Ripon were bought at 105 and 45 guineas respectively, as well as some of the choicest Ewes, at prices as high as twelve and fifteen guineas each.

Many of the most famous Sheep in England have sprung from these animals, viz: Royal Newcastle, the Jubilee Champion, Baronet, Royal Plymouth, and others.

Since that time Sheep from the Throckmorton, Penfold, Coleman, and Webb flocks have been introduced when a change of blood was required.

Numerous prizes have been taken during the past seven years, including highest honours, both at the breeding and fat-stock shows. On the Continent, at the International Show at Paris in 1889, three champion prizes were awarded, as also the Prix d'Honneur for the best collection of Sheep of any variety.

The Sheep have been equally successful when shown at the American and Canadian Exhibitions.

Ewes 320   Rams 23
Shearling Ewes ... 155

FLOCK No. 13.

THE DUKE OF NORTHUMBERLAND, Albury, Guildford.

Agent's Address—A. PEEBLES, Albury.

This flock was established in 1877, and fresh blood has been introduced from time to time by purchasing Ewes from the flocks of Sir William Throckmorton, Bart., and Mr. Botting.

The Rams used have been selected with great care from the flocks of Messrs. J. J. Colman, Carew Gibson, Edwin Ellis, A. Heasman, H. Penfold, W. Toop, and Lord Onslow.

Ewes 300   Rams 40
Shearling Ewes 100

FLOCK No. 14.

W. FLUX, Ampney Crucis, Cirencester.

This flock was founded by a draft of Ewes bought from Earl Bathurst, and Rams have been used from the flocks of Sir W. Throckmorton, Bart., and Messrs. Penfold and Heasman.

Ewes 44   Rams 9
Shearling Ewes 42

FLOCK No. 15.

EARL BATHURST, Cirencester.

Agent's Address—ROBT. ANDERSON, The Bartons, Cirencester.

The Cirencester Park Flock was founded very early in the century, purchases being made from Mr. George Talbot, who for several years took the whole of the draft Ewes from Mr. John Ellman, of Glynde, and brought them to his estate in Gloucestershire. As early as 1820 and subsequently Rams and Ewes were procured from Mr. Ellman, the Duke of Richmond and other noted Sussex breeders. In 1844 the Babraham blood was introduced by a purchase from Mr. Jonas Webb. Of recent years Rams have been selected from the celebrated flocks of the Duke of Richmond, Sir William Throckmorton, Messrs. William Rigden, J. J. Colman, C. Chapman, Henry Webb, William Toop and Edwin Ellis.

Ewes 554   Rams 11
Shearling Ewes 190

FLOCK No. 16.

HERBERT PADWICK, Thorney Island, Emsworth.

This flock, originally formed by Mr. David Padwick, has traditions inferior to none in the kingdom. Being formed in 1820 from the best Ewes which could be procured, it was an established flock in 1830 when it passed into the hands of Mr. Frederick Padwick, son of Mr. David Padwick, who at first used Rams from the flocks of the Brothers Ellman, the Duke of Richmond, and Messrs. Atwick Pinnix and H. P. Hart, and who added to it by purchases made at Mr. Robert Drewitt's sale (August, 1847), whose flock was at that time considered the finest in existence. Mr. Padwick, being a relative of that celebrated Southdown breeder, Mr. Henry Percival Hart, of Beddingham, near Lewes, from the year 1846, bred almost exclusively from his Rams—so much was this the case that if other Rams were occasionally used their produce was invariably drafted, and recourse was again had to the Beddingham flock. It will be remembered that Mr. Hart was from childhood an intimate friend and near neighbour of the Ellmans, who have been called the founders of the "Improved Southdown"—indeed, his farm lay immediately between those of Messrs. John and Thomas Ellman, and he thus possessed facilities for studying Southdown, shared, perhaps, by no other breeder. Early taking an interest in sheep-breeding, he obtained much information from the Ellmans, and also from their shepherd, to whom John Ellman attributed much of his success. Possessing, then, this knowledge and these advantages, Mr. Hart was enabled, on the dispersion of Mr. John Ellman's flock, to bring together into his own the best strains of "Ellman" blood, and, in a sense, to perpetuate the founder's flock.

The Thorney Island flock being so closely related to his own, Mr. Hart has naturally always taken a great interest in it, an interest which became intensified upon his ceasing for a time, after his sale in 1878, to be himself a flock-master. Since then he has given the present owner (Mr. Herbert Padwick) the assistance of his great experience and judgment in the selection of Rams to replenish the Thorney flock, an assistance which this opportunity enables the owner gratefully to acknowledge. The "Hart blood" is now so generally known and so highly valued, that it is unnecessary to do more than give the number of animals bought or hired from the Beddingham flock, and to mention that the use of Ram Lambs has been studiously avoided—a point upon which Mr. Hart has always laid the greatest stress. Hart Rams were used yearly from 1847 to 1878 (five were bought in 1847). During that period 40 were bought or hired, and also 220 Ewes ; and as lately as 1889 a pure Hart Ram was in use. The number kept in the flock was 300 from 1820 to 1846, 350 to 1878. Since the latter year the flock has been gradually raised to 500 (its present number), the greatest increase taking place when 70 Ewes were added from the Goodwood flock.

Ewes	...	...	...	470	Rams	...	...	...	13
Ewe tegs ...		...	...	140			

FLOCK No. 17.

J. W. BAXENDALE, Hursley Park, Winchester.

Ewes	...	...	...	135	Rams	...	...	...	12
Shearling Ewes ...		...	50			

FLOCK No. 18.

JAMES BLYTH, Blythwood, Stanstead, Essex.

The Ewes of this flock are chiefly Ewes bred by H. Webb, of Streetly Hall, Cambridge, and descended from the following Rams :—Gloucester, General Favourite, Guinea Peru, Little Twin, Melton, Foljambe, No. 8 in 1885, Derby,

No. 1 in 1884, Handsome Head, Lamb by Clipstone, Hardihood, St. Blaize, No. 20 in 1883, Peregrine, Foljame's Favourite, Captain, and Shepherds' Favourite, large purchases having been made at both of the sales of Mr. Webb, and Rams from the celebrated flock of J. J. Colman, M.P., have been hired and used in this flock and also a selection of Ewes.

The flock has in this year, 1892, been very successful at the Royal and County shows.

Flock Ewes	...	...	170	Rams ...	... ...	14
Shearling Ewes ...	...	52				

FLOCK No. 19.

Mrs. PERKINS, Saham Hall, Wotton, Norfolk.

The Saham Hall flock has been in existence many years, and it has always been kept replenished with purchases from the leading flocks of the country. It was strengthened very largely at the sales taking place at Merton on the dispersion of Lord Walsingham's flock, and again at Mr. Henry Webb's sales, and great care has been taken in connection with the pedigrees of the flock; and in addition to the entry of Rams in the Flock Book thirty individual Ewes of Merton blood have been entered, and these are being crossed with Webb's Rams to raise a flock of Sheep, the pedigrees of which will all be registered.

Ewes	...	...	... 240	Rams ...	... ...	5
Shearling Ewes ...	...	60				

FLOCK No. 20.

Messrs. PAGE-WOOD & SON, Wakes Hall, Wakes Colne, Essex.

This flock was founded by the purchase of 80 Ewes from Mr. W. Everitt, of Brightwell, Norfolk, and replenished by Ewes purchased at the Marquis of Bristol's sales, 1891.

Rams used in the flock have been purchased from the Marquis of Bristol and Mr. Henry Webb.

Old Ewes...	...	... 111	Old Rams ...	...	... 4
Shearling Ewes ...	...	46	Shearling Rams	...	... 3

FLOCK No. 21.

WALTER WOODS, Langford, near Chichester.

This old-established flock has its habitat on the Sussex Downs, and the present owner has for many years exercised the greatest care in the selection of Rams, which have been purchased from Mr. Penfold and other well known breeders.

Old Ewes...	...	... 440	Rams	... ...	... 12
Shearling Ewes ...	...	210				

FLOCK No. 22.

H. M. SIMMONS, St. Leonards Villa, Eastbourne and Jevington Farm, Eastbourne.

This is a hard-working Sussex Hill flock, and great care has been taken in the selection of Rams. Of late years Rams from H.R.H. Prince of Wales, Henry Webb, and Hugh Penfold have been used.

Ewes	...	...	... 800	Rams	... ...	14

FLOCK No. 23.

LAURENCE CAVE, Ditcham Park, Petersfield.

Agent's Address—EDWIN HILL, Old Ditcham, Petersfield.

Ewes 231

FLOCK No. 24.

C. CZARNIKOW, Effingham Hill, Dorking, Surrey.

This flock has been founded by Ewes bought chiefly from Lord Onslow and Sir Wm. Throckmorton, and Rams have been used from the flocks of E. Ellis, Newman, &c.

Old Ewes 50   Shearling Ewes ... 25

FLOCK No. 25.

T. BANNISTER, Hayward's Heath, Sussex.

This flock was founded by Ewes purchased at the sale of Mr. Stuart's flock at Ardingly, which contained Sheep of the best blood only, and consisted very largely of Heasman Ewes obtained at the Calceto Sale in 1889.

Ewes 46

FLOCK No. 26.

HENRY JAY, Fitzleroi Farm, Fittleworth, Sussex.

This flock is a hard-working Sussex flock, and the owner has taken great care to use Rams of the best strains.

Ewes 300   Old Ram 1
                        Shearling Rams 5

FLOCK No. 27.

DAVID WADDINGTON, Manor Farm, Lavant, Chichester.

This flock was founded in 1888 by purchase of Ewes from the Duke of Richmond, Messrs. Mannington and Drewitt, and Rams have been used from the flocks of Messrs. Toop, Hampton, Heasman, and Newman.

Ewes 485   Rams 6

RAMS AND RAM LAMBS.

Entered by Messrs. DE MURIETTA, Wadhurst Park, Wadhurst, Sussex.

FLOCK NO. 1.

(N.B.—Numbers in brackets ( ) are ear numbers. All ear numbers are in right ear unless otherwise stated.)

Flock
Book No.

1 Ram (91) of 1890.
  Sire No. 52 in Flock Book.   Dam, (327 left) Webb Ewe.
                  Sire, Webb's Derby. Dam by Baltimore 2nd.

2 Ram (92) of 1890.
  Sire No. 51 in Flock Book.   Dam (100 left), Penfold Ewe.
                  Sire, Brighton Champion, 1885.

3 Ram (96) of 1890.
  Sire No. 48 in Flock Book.   Dam (366 left), Sandringham Ewe.
                  1st Sire, a Walsingham Ram. Dam, Sandringham Ewe.
                  2nd Sire, Royal Shrewsbury.

4 Ram (97) of 1890.
  Sire No. 51 in Flock Book.   Dam (100 left), Penfold Ewe.
                  Sire, Brighton Champion, 1885.

5 Ram (105) of 1890.
  Sire No. 49 in Flock Book.   Dam (36 left), Gibson Ewe.

6 Ram (109) of 1390.
  Sire No. 52 in Flock Book.   Dam (316 left), Webb Ewe.
                  Sire, Webb's Ormonde. Dam by No. 18 in 1876.

7 Ram (138) of 1890.
  Sire No. 52 in Flock Book.   Dam (311 left), Webb Ewe.
                  Sire, Delorme. Dam by Guinea Peru.

8 Ram (216) of 1891.
  Sire No. 290 in Flock Book.   Dam (45 left), Kingscote Ewe.

9 Ram (246) of 1891.
  Sire No. 290 in Flock Book.   Dam (329 left), Webb Ewe.
                  Sire, Webb's Derby. Dam by No. 31 in 1875.

10   Ram (249) of 1891.
     Sire No. 290 in Flock Book.   Dam (420 left).
          1st Sire, a Colman Ram.   Dam, Colman Ewe.
          2nd Sire, Colman's Old Norwich. Dam by Webb's Hardihood.
          3rd Sire, Webb's Derby.

11   Ram (251) of 1891.
     Sire No. 290 in Flock Book.   Dam (415 left).
          1st Sire, Colman's No. 5 Ram Lamb. Dam, a Colman Ewe.
          2nd Sire, Lt. Guinea Peru 2nd.
          3rd Sire, Webb's 20 in 1883. Dam by Webb's Streetley.

12   Ram (200) of 1891.
     Sire No. 52 in Flock Book.   Dam (250 left), Kingscote Ewe.
          Sire, a Duke of Richmond Ram.

13   Ram (207) of 1891.
     Sire No. 52 in Flock Book.   Dam (22 left), Kingscote Ewe.
          Sire, a Walsingham Ram.

14   Ram (213) of 1891.
     Sire No. 52 in Flock Book.   Dam (259 left), Kingscote Ewe.
          Sire, a Chapman Ram.

15   Ram (214) of 1891.
     Sire No. 52 in Flock Book.   Dam, a Ewe, bred by Mr. C. G. Carew Gibson.

16   Ram (215) of 1891.
     Sire No. 52 in Flock Book.   Dam (100 left), Penfold Ewe.
          Sire, Brighton Champion, 1885.

17   Ram (217) of 1891.
     Sire No. 52 in Flock Book.   Dam (247 left), Kingscote Ewe.
          Sire, a Duke of Richmond Ram.

18   Ram (218) of 1891.
     Sire No. 52 in Flock Book.   Dam (72 left), Colman Ewe.
          Sire, a Throckmorton Ram.

19   Ram (220) of 1891.
     Sire No. 52 in Flock Book.   Dam (87 left), Penfold Ewe.
          Sire, a Hart Ram.

20   Ram (223) of 1891.
     Sire No. 52 in Flock Book.   Dam (97 left), Penfold Ewe.
          Sire, Brighton Champion, 1885.

21   Ram (228) of 1891.
     Sire No. 52 in Flock Book.   Dam (118 left), Kingscote Ewe.
          Sire, a Walsingham Ram.

22   Ram (233) of 1891.
     Sire No. 52 in Flock Book.   Dam (184 left), a Gibson Ewe.

23   Ram (239) of 1891.
     Sire No. 52 in Flock Book.   Dam (445 left), Wadhurst Park Ewe.
                                         1st Sire, No. 30, Gibson Sale 1887.
                                         Dam, Gibson Ewe.
                                         2nd Sire, a Grandson of a Rigden
                                         Ram. Dam, Rigden Ewe.

24   Ram (240) of 1891.
     Sire No. 52 in Flock Book.   Dam (31 left), a Gibson Ewe.

25   Ram (243) of 1891.
     Sire No. 52 in Flock Book.   Dam (446 left), Wadhurst Park Ewe.
                                           1st Sire, No. 30 of Gibson's Sale 1887.
                                         Dam, a Gibson Ewe.
                                         2nd Sire, a Grandson of a Rigden
                                         Ram. Dam, a Rigden Ewe.

26   Ram (255) of 1891.
     Sire No. 290 in Flock Book.   Dam (326 left), Webb Ewe.
                                           Sire, Derby.

27   Ram (201) of 1891.
     Sire No. 54 in Flock Book.   Dam (278 left), Kingscote Ewe.

28   Ram (208) of 1891.
     Sire No. 54 in Flock Book.   Dam (108 left), Penfold Ewe.
                                           1st Sire, Son of Brighton Champion,
                                         1885. Dam, a Penfold Ewe.
                                         2nd Sire, Brighton Champion, 1885.

29   Ram (209) of 1891.
     Sire No. 54 in Flock Book.   Dam (393) left, Colman Ewe.
                                           1st Sire, Colman's Gloucester.
                                           2nd Sire, Webb's Gloucester.
                                           3rd Sire, Webb's General Favourite.

30   Ram (210) of 1891.
     Sire No. 54 in Flock Book.   Dam (316 left), Webb Ewe.
                                           Sire, Webb's Ormonde. Dam by No.
                                           18 in 1876.

31   Ram (211) of 1891.
     Sire No. 54 in Flock Book.   Dam (49 left), Kingscote Ewe.
                                           Sire, a Bathurst Ram.

32   Ram (226) of 1891.
     Sire No. 54 in Flock Book.   Dam (402 left).
                                           1st Sire, Colman's No. 5 Ram Lamb,
                                         1886. Dam, a Colman Ewe.
                                         2nd Sire, Lt. Guinea Peru 2nd
                                         (Webb's).
                                         3rd Sire, Webb's, 20 in 1883.

Flock
Book No.

33 Ram (230) of 1891.
   Sire No. 54 in Flock Book. Dam (312 left), Webb Ewe.
   Sire, Webb's Delorme. Dam by Shepherds' Favourite.

34 Ram (231) of 1891.
   Sire No. 54 in Flock Book. Dam (24 left), Colman Ewe.
   Sire, a Throckmorton Ram. Dam, a Colman Ewe.

35 Ram (234) of 1891.
   Sire No. 54 in Flock Book. Dam (445 left), Wadhurst Park Ewe.
   1st Sire, No. 30 of Gibson Sale, 1887. Dam, a Gibson Ewe.
   2nd Sire, a Grandson of a Rigden Ram. Dam, a Rigden Ewe.

36 Ram (237) of 1891.
   Sire No. 54 in Flock Book. A Dam (405 left), Colman Ewe.
   1st Sire, Colman's No. 5 Ram Lamb, 1886. Dam, a Colman Ewe.
   2nd Sire, Webb's Lt. Guinea Peru 2nd.
   3rd Sire, Webb's 20 in 1883.

37 Ram (245) of 1891.
   Sire No. 54 in Flock Book. Dam (402 left), Colman Ewe.
   1st Sire, Colman's No. 5 Ram Lamb, 1886. Dam, a Colman Ewe.
   2nd Sire, Webb's Lt. Guinea Peru.
   3rd Sire, Webb's 20 in 1883.

38 Ram (221) of 1891.
   Sire No. 53 in Flock Book. Dam (378 left), Colman Ewe.
   1st Sire, Webb's 62 of 1887. Dam, a Colman Ewe.
   2nd Sire, Webb's Gloucester.

39 Ram (236) of 1891.
   Sire No. 53 in Flock Book. Dam (293 left), Kingscote Ewe.
   Sire, a Walsingham Ram.

40 Ram (247) of 1891.
   Sire No. 53 in Flock Book. Dam (390 left), Colman Ewe.
   1st Sire, Webb's 62 of 1887. Dam, a Colman Ewe.
   2nd Sire, Webb's Gloucester.

41 Ram (254) of 1891.
   Sire No. 53 in Flock Book. Dam (90 left), Penfold Ewe.
   Sire, Brighton Champion, 1885.

42 Ram Lamb (261) of 1892.
   Sire No. 215 in Flock Book. Dam (450 left), a Wadhurst Park Ewe.
   1st Sire, No. 30 of Gibson's Sale, 1887. Dam, a Gibson Ewe.
   2nd Sire, a Grandson of a Rigden Ram. Dam, a Rigden Ewe.

Flock
Book No.

43 Ram Lamb (267) of 1892.
    Sire No. 215 in Flock Book.   Dam (23 left), Kingscote Ewe.
                           Sire, Walsingham Ram.

44 Ram Lamb (271) of 1892.
    Sire No. 215 in Flock Book.   Dam (243 left), Kingscote Ewe.
                           Sire, a Walsingham Ram.

45 Ram Lamb (275) of 1892.
    Sire No. 215 in Flock Book.   Dam (93 left), Penfold Ewe.
                           Sire, Brighton Champion, 1885.

46 Ram Lamb (278) of 1892.
    Sire No. 58 in Flock Book.   Dam (411 left), Colman Ewe.
                          1st Sire, Colman's No. 5 Ram Lamb, 1886.
                          2nd Sire, Webb's Lt. Guinea Peru.
                          3rd Sire, Webb's Gloucester.

47 Ram Lamb (309) of 1892.
    Sire No. 57 in Flock Book.   Dam (473 left), Webb Ewe.
                          Sire, Webb's Gloucester. Dam by No. 1 in 1884.

48 Ram, born 1887. Penfold No. 24 of 1888.
    Sire, a son of H. Webb's   Dam, a Penfold Ewe.
      General   Favourite,   Sire, Hart Ram.
      Dam, Penfold Ewe.
    2nd Sire, Webb's General
      Favourite. Dam by
      Flack's Favourite.
    3rd Sire, No. 20 in 1888.

49 Ram born 1887. Duke of Richmond's No. 7 of 1888.
    1st Sire, a Ram bred by Mr.   Dam, a Goodwood Ewe.
      Chapman. Dam, Chap-
      man Ewe.
    2nd Sire, a son of Duke of
      York. Dam, Chapman
      Ewe.
    3rd Sire, Duke of York.
      Dam, Chapman Ewe.
    4th Sire, Buckland. Dam,
      Throckmorton Ewe.

50 Ram, born 1888. No. 1 Sandringham Sale, 1889.
    Sire, Portsmouth.   Dam, a Sandringham Ewe.
                             Sire, Carlisle.

51 Ram, born 1883. Gloucester. Lot 2, Webb's Sale, 1889.
    1st Sire, General Favourite. Dam, a Webb Ewe.
      Dam   by   Flack's   Sire, No. 38 of 1875.
      Favourite.
    2nd Sire, Webb's No. 20 in
      1880.

52 Ram born 1888. Webb's No. 39 of 1889 Sale.
     1st Sire, Enterprise. Dam   Dam, a Webb Ewe.
      by Norfolk Favourite.   Sire, Handsome Head.
     2nd Sire, Gloucester. Dam
      by No. 38 of 1875.
     3rd Sire, General Favourite.
      Dam by Flack's
      Favourite.
     4th Sire, No. 20 in 1880.

53 Ram. No. 10. Ram Lamb of Calceto Sale, 1889.
     1st Sire, Brighton.   Dam, a Heasman Ewe.
     2nd Sire, Heasman Ram.
     3rd Sire, Penfold Ram.

54 Ram born 1889. Webb's No. 44 of 1890 Sale.
     1st Sire, Cambridgeshire. Dam, Webb Ewe.
      Dam by St. Blaize.   Sire, Webb's No. 53 in 1880.
     2nd Sire, Gloucester. Dam
      by No. 38 in 1875.
     3rd Sire, General Favourite.
      Dam by Flack's
      Favourite.
     4th Sire, No. 20 in 1880.

55 Ram born 1888. No. 28 of Throckmorton's Sale 1890.
     1st Sire, Webb's 23 in 1888. Dam, a Throckmorton Ewe.
      Dam by Guinea Peru.
     2nd Sire, Webb's Hardihood.
      Dam by Throckmor-
      ton's Favourite.
     3rd Sire, Webb's Baltimore
      1st.

56 Ram. Ellis' No. 75 of 1890.
     1st Sire, Baronet. Dam, a   Dam, a Walsingham Ewe.
      Throckmorton Ewe.
     2nd Sire, Ripon. Dam, a
      Ewe by Son of Royal
      Taunton.
     3rd Sire, Royal Reading.
      Dam, a Walsingham
      Ewe.

57 Ram born 1889. Throckmorton No. 1, 1891 Sale.
     1st Sire, Nottingham. Dam, Dam, a Throckmorton Ewe.
      a Goodwood Ewe.
     2nd Sire, a Grandson of
      Duke of York. Dam,
      a Goodwood Ewe.
     3rd Sire, a Son of Duke of
      York. Dam, a Chap-
      man Ewe.
     4th Sire, The Duke of York
      Dam, a Chapman Ewe.

Flock
Book No.

58 Ram. Ellis' No. 389 of 1891.
 1st Sire, Royal Plymouth. Dam, an Ellis Ewe.
  Dam, a Walsingham 1st Sire, Colman's No. 18 out of
  Ewe.   Colman Ewe.
 2nd Sire, Baronet. Dam, a 2nd Sire, Webb's No. 47.
  Throckmorton Ewe.
 3rd Sire, Ripon. Dam, a
  Ewe by Son of Royal
  Taunton.
 4th Sire, Royal Reading.
  Dam, a Walsingham
  Ewe.

59 Ram born 1890. No. 63 of Chichester Sale, 1891.
 1st Sire, Favourite. Dam, a Heasman Ewe.
 2nd Sire, Brother to
  Brighton.
 3rd Sire, Heasman Ram.
 4th Sire, a Penfold Ram.

60 Ram born 1889. No. 44 of Chichester Sale, 1891.
 1st Sire, a Son of Rigden's Dam, an Emery Ewe.
  Royal Derby.
 2nd Sire, Rigden's Royal
  Derby.

(*Continued on page* 50.)

Entered by THE PAGHAM HARBOUR COMPANY, Selsey, Chichester.

(*Numbers in brackets ( ) are ear numbers, and are in the right ear.*)

FLOCK No. 3.

61 "**King William**" (61), born in 1882.
 Sire bred by Charles Hobgen. Dam, Pagham Harbour Ewe,
 2nd Sire, Penfold Ram. by Penfold Ram.

62 "**Royal Nottingham**" (62), born in 1888.
 Sire, Selsey, No. 292 in Flock
  Book.   Dam, Pagham Harbour Ewe,
 2nd Sire, Webb's General Favourite. by 61 in Flock Book.
 3rd Sire, Webb's No. 20 in 1880.

63 "**Beacon**" (63), born in 1886.
 (Penfold's No. 37 in 1888 sale.) Dam, Penfold Ewe.
 Sire, Penfold's Young Crawford.
 2nd Sire, Penfold's Crawford.
 3rd Sire, Penfold's No. 22 in 1875.
 4th Sire, Ram Lamb, bred by
  Lord Walsingham, No. 49 in
  1871 sale, No. 343 in Flock
  Book.

64 "**Windsor**" (64), born in 1888.
 Sire, Selsey, bred by Penfold, Dam, Penfold Ewe, by Hart
  No. 292 in Flock Book. Ram.

65 **"Windsor 2nd"** (65), born in 1890.
    Sire, Windsor, No. 64 in Flock   Dam, Goodwood Ewe.
    Book.

66 **"Webb"** (66), born in 1889.
    Sire, Selsey, No. 292 in Flock   Dam, Pagham Harbour Ewe,
    Book.   by Son of 61

67 **"Royal Nottingham 2nd"** (67), born
    in 1890.
    Sire, Royal Nottingham, No. 62   Dam, Pagham Harbour Ewe,
    in Flock Book.   by Son of 61.

68 **"Champion 3rd"** (68), born in 1889.   Dam, Pagham Harbour Ewe.
    By Champion 2nd, No. 297 in
    Flock Book.

69 **"Champion 4th"** (69), born in 1889.   Dam, Pagham Harbour Ewe,
    By Champion 2nd, No. 297 in   By 61 in Flock Book.
    Flock Book.

70 **"Penfold"** (70), No. 177, Penfold's 1891   Dam, Penfold Ewe.
    Chichester Sale (70), born in 1889.
    Sire, Duke of Richmond's No. 10
    in 1888, out of a Rigden Ewe.
    2nd Sire, a Webb Ram.

71 **"Beacon 2nd"** (71), born in 1889. By   Dam, Goodwood Ewe.
    No. 63 in Flock Book.

72 **"Windsor 4th"** (72), born in 1890. By   Dam, Penfold Ewe.
    No. 64 in Flock Book.

73 **"Champion 6th"** (73), born in 1891.   Dam, Pagham Harbour Ewe,
    By No. 297 in Flock Book.   by 62 in Flock Book.

74 **"Selsey Bill 2nd"** (74), born in 1891.   Dam, Pagham Harbour Ewe.
    By No. 102 in Flock Book.   By Son of 61 in Flock
                          Book.

75 **"Selsey Bill 3rd"** (75), born in 1891.   Dam, Pagham Harbour Ewe.
    By No. 102 in Flock Book.   By Son of 61 in Flock Book.

76 **"Selsey Bill 4th"** (76), born in 1891.   Dam, Pagham Harbour Ewe.
    By No. 102 in Flock Book.

77 **"Cambridgeshire 4th"** (77), born in   Dam, bred by Mr. Ellis, by
    1891. By No. 80 in Flock Book.   Merton, out of Throck-
                          morton Ewe.

78 Ram, **"Little Cambridgeshire"** (78),   Dam, Pagham Harbour Ewe.
    born in 1891. By No. 80 in
    Flock Book.

79 Ram (79), born in 1891. By No. 71   Dam, Goodwood Ewe.
    in Flock Book.

80   " **Cambridgeshire 2nd** " (80), born in
1889. No. 82 in Webb's 1890
Sale.   Dam, Webb Ewe. Sire, No. 15
Sire, Cambridgeshire. Dam by St.   in 1886.
   Blaize.
2nd Sire, Gloucester. Dam by 38 in
   1875.
3rd Sire, General Favourite. Dam
   by Flack's Favourite.
4th Sire, Webb's No. 20 in 1880.

81   " **Windsor 3rd** " (81), born in 1890.
by No. 64 in Flock Book.   Dam, Penfold Ewe.

82   " **Hardy** " (82), No. 10 in Ellis Sale,
1887, born in 1886.   Dam, Throckmorton Ewe.
Sire, Merton. Dam, a Ewe by
   Grandson of Royal Taunton.

83   " **Enterprise 2nd** " (83), born in 1889.
No. 87 in Webb's 1890 Sale.   Dam, Webb Ewe by Melton.
Sire, Enterprise.
2nd Sire, Gloucester. Dam by 38
   in 1875.
3rd Sire, General Favourite.
4th Sire, Webb's No. 20 in 1880.

84   " **King William 2nd** " (84), born in
1889, by No. 61 in Flock
Book.   Dam, Pagham Harbour Ewe,
     by 292 in Flock Book.

85   " **Champion 5th** " (85), born in 1890,
by No. 297 in Flock Book.   Dam, Pagham Harbour Ewe,
     by 61 in Flock Book.

86   " **Cambridgeshire 3rd** " (86), born in
1891, by No. 80 in Flock
Book.   Dam, Pagham Harbour Ewe,
     by 292 in Flock Book.

87   " **Royal Nottingham 3rd** " (87), born
in 1891, by No. 62 in Flock
Book.   Dam, Pagham Harbour Ewe,
     by Penfold Ram. No.
     297 in Flock Book.

88   " **Royal Nottingham 4th** " (88), born
in 1891, by No. 62 in Flock Book.   Dam, Pagham Harbour Ewe.

89   Ram (89), born in 1891, by No. 80 in
Flock Book.   Dam, Pagham Harbour Ewe.

90   Ram (90), born in 1891, by No. 69 in
Flock Book.   Dam, Pagham Harbour Ewe.

91   Ram (91), born in 1891, by No. 69 in
Flock Book.   Dam, Pagham Harbour Ewe.

Flock
Book No.

92 Ram (92), born in 1891, by No. 69 in
Flock Book.   Dam, Pagham Harbour Ewe,
                                                by 291 in Flock Book.

93 Ram (93), born in 1891, by No. 69 in
Flock Book.   Dam, Pagham Harbour Ewe.

94 Ram (94), born in 1891, by No. 102 in
Flock Book.   Dam, Pagham Harbour Ewe.

95 Ram (95), born in 1891. Bred by Mr.
Stuart. No 16 at his sale.   Dam, Heasman Ewe.
Sire, Penfold Ram. No. 84 in
1888.
2nd Sire, Penfold's Brighton
Champion, 1885.

96 Ram (96), born in 1891.   Dam, Pagham Harbour Ewe.
Sire, 291 in Flock Book.

97 Ram (97), born in 1891.
Sire, No. 69 in Flock Book.   Dam, Pagham Harbour Ewe.

98 Ram (98), born in 1891.
Sire, No. 80 in Flock Book.   Dam, Pagham Harbour Ewe.

99 Ram (99), born in 1891.
Sire, No. 80 in Flock Book.   Dam, Pagham Harbour Ewe.

100 Ram, born in 1890. Sire No. 62 in
Flock Book.   Dam, a Pagham Harbour Ewe.

(*Continued on page* 51.)

Entered by HUGH PENFOLD, Selsey, Chichester.
FLOCK No. 4.

101 Ram, born in 1888, " **Constitution.**"
Sire, a son of Webb's General   Dam, a Penfold Ewe. Sire,
Favourite.   Hart Ram. Dam, a Pen-
2nd Sire, Webb's General Favourite.   fold Ewe.
Dam, by Flack's Favourite.
3rd Sire, Webb's No. 20 in 1880.

102 Ram, born in 1889, " **Selsey Bill.**"
Sire, Duke of Richmond's No. 10   Dam, a Penfold Ewe.
of 1888 out of a Rigden Ewe.
2nd Sire, a Webb Ram.

103 Ram, born in 1887, Gibson's No. 1 of
1890 sale.
Sire, a Gibson Ram.   Dam, a Gibson Ewe.
2nd Sire, Mr. Rigden's Ram.
3rd Sire, a Webb Ram.

Flock
Book No.

104 Ram, born in 1888, "**Norton,**" by
No. 63 in Flock Book.
Dam, a Penfold Ewe.

105 Ram, born in 1889, "**Sussex.**"
Sire, Duke of Richmond's No. 10
of 1888, out of a Rigden Ewe.
Dam, a Penfold Ewe.
2nd Sire, a Webb Ram.

106 Ram, born in 1890, "**Chichester.**"
Sire, Goodwood.
2nd Sire, Duke of Richmond's
No. 10 of 1888, out of a
Rigden Ewe.
Dam, a Penfold Ewe.
3rd Sire, a Webb Ram.

107 Ram, born in 1890, "**Selsey Prince.**"
Sire, No. 101 in Flock Book.
Dam, a Penfold Ewe.

108 Ram, born in 1890, "**Selsey Boy.**"
Sire, No. 101 in Flock Book.
Dam, a Penfold Ewe.

109 Ram, born in 1890.
Sire, Goodwood.
2nd Sire, Duke of Richmond's
No. 10 of 1888, out of a
Rigden Ewe.
Dam, a Penfold Ewe.
3rd Sire, a Webb Ram.

110 Ram, born in 1890.
Sire, Goodwood.
2nd Sire, Duke of Richmond's
No. 10 of 1888, out of a
Rigden Ewe.
Dam, a Penfold Ewe.
3rd Sire, a Webb Ram.

111 Ram, born in 1890.
Sire, Goodwood.
2nd Sire, Duke of Richmond's
No. 10 of 1888, out of a
Rigden Ewe.
Dam, a Penfold Ewe.
3rd Sire, Webb Ram.

112 Ram, born in 1890, "**Selsey Lad.**"
Sire, No. 101 in Flock Book.
Dam, a Penfold Ewe.

113 Ram, born in 1890.
Sire, No. 101 in Flock Book.
Dam, a Penfold Ewe.

114 Ram, born in 1890, "**Selsey
Champion.**"
Sire, Son of Brighton Champion,
1885. No. 341 in Flock Book.
Dam, a Penfold Ewe.

115 Ram, born in 1891.
Sire, No. 101 in Flock Book.
Dam, a Penfold Ewe.

116 Ram, born in 1891.
     Sire, No. 101 in Flock Book.   Dam, a Penfold Ewe.

117 Ram, born in 1891.
     Sire, No. 101 in Flock Book.   Dam, a Penfold Ewe, by Son
                                        of Webb's "Quality."

118 Ram, born in 1891.
     Sire, No. 101 in Flock Book.   Dam, a Penfold Ewe.

119 Ram, born in 1891.
     Sire, No. 105 in Flock Book.   Dam, a Penfold Ewe.

120 Ram, born in 1891.
     Sire, No. 105 in Flock Book.   Dam, a Penfold Ewe.

121 Ram, born in 1891.
     Sire, No. 105 in Flock Book.   Dam, a Penfold Ewe.

122 Ram, born in 1891.
     Sire, No. 105 in Flock Book.   Dam, a Penfold Ewe.

123 Ram, born in 1891.
     Sire, No. 105 in Flock Book.   Dam, a Penfold Ewe.

124 Ram, born in 1891.
     Sire, No. 105 in Flock Book.   Dam, a Penfold Ewe.

125 Ram, born in 1891.
     Sire, No. 105 in Flock Book.   Dam, a Penfold Ewe.

126 Ram, born in 1891.
     Sire, No. 105 in Flock Book.   Dam, a Penfold Ewe.

127 Ram, born in 1891.
     Sire, No. 175, Chichester Sale,
        1891.   Dam, a Penfold Ewe.
     2nd Sire, Duke of Richmond's No.
        10 of 1888, out of a Rigden
        Ewe.
     3rd Sire, a Webb Ram.

128 Ram, born in 1891.
     Sire, No. 175, Chichester Sale,
        1891.   Dam, a Penfold Ewe.
     2nd Sire, Duke of Richmond's No.
        10 of 1888, out of a Rigden
        Ewe.
     3rd Sire, a Webb Ram.

129 Ram, born in 1891.
     Sire, No. 175, Chichester Sale,
        1891.   Dam, a Penfold Ewe.
     2nd Sire, Duke of Richmond's No.
        10 of 1888, out of a Rigden
        Ewe.
     3rd Sire, a Webb Ram.

Flock
Bock No.

130 Ram, born in 1891.
Sire, No. 175, Chichester Sale,
1891.   Dam, a Penfold Ewe.
2nd Sire, Duke of Richmond's No.
10 of 1888, out of a Rigden
Ewe.
3rd Sire, a Webb Ram.

131 Ram, born in 1891.
Sire No. 80 in Flock Book.   Dam, a Penfold Ewe.

132 Ram Lamb, born in 1892.
Sire, No. 82 in Flock Book.   Dam, a Penfold Ewe.

133 Ram Lamb, born in 1892.
Sire, No. 80 in Flock Book.   Dam, a Penfold Ewe.

134 Ram Lamb, born in 1892.
Sire, No. 80 in Flock Book.   Dam, a Penfold Ewe.

135 Ram Lamb, born in 1892.
Sire, No. 102 in Flock Book.   Dam, a Penfold Ewe.

136 Ram, born in 1891, "**Selsey King**."
1st Sire, Son of Cambridgeshire,
out of Woodman Ewe.   Dam, a Penfold Ewe.
2nd Sire, Webb's Cambridgeshire.
Dam by St. Blaize.
3rd Sire, Webb's Gloucester. Dam,
by No. 38 in 1875.
4th Sire, General Favourite
(Webb's), by Flack's Favourite.

(*Continued on page* 55.)

Entered by FRANCIS NEALE HOBGEN, Appledram, Chichester.

FLOCK No. 6.

137 Ram, born in 1889. Bred by Lord
Alington.

138 Ram, born in 1889.
Sire, a Ram bred by Mr. Edward
Hobgen.

139 Ram, born in 1891.
Sire, Penfold, No. 120 in 1888.
2nd Sire, Son of Brighton Champion, 1885.
3rd Sire, Brighton Champion, 1885.

140 Ram, born in 1891.
1st Sire, Penfold's 120 in 1888.
2nd Sire, Son of Brighton
Champion, 1885.
3rd Sire, Brighton Champion,
1885.

141 Ram, born in 1891.
      Sire, No. 62 in Flock Book.

142 Ram, born in 1891.
      Sire, No. 62 in Flock Book.

143 Ram, born in 1891.
      Sire, No. 62 in Flock Book.

144 Ram, born in 1891.
      Sire, No. 62 in Flock Book.

145 Ram, born in 1891.
      Sire, No. 62 in Flock Book.

146 Ram, born in 1891.
      Sire, No. 62 in Flock Book.

147 Ram, born in 1891.
      Sire, No. 62 in Flock Book.

148 Ram, born in 1891.
      Sire, No. 62 in Flock Book.

149 Ram, born in 1891.
      Sire, No. 62 in Flock Book.

150 Ram, born in 1891.
      Sire, No. 62 in Flock Book.

151 Ram, born in 1891.
      Sire, No. 62 in Flock Book.

152 Ram, born in 1891.
      Sire, a Son of E. Ellis'
         Marquis.
      2nd Sire, Marquis.

153 Ram, born in 1891.
      Sire, a Son of Merton.
      2nd Sire, Merton, Lord Walsing-
         ham's.

154 Ram, born in 1891.
      Sire, a Son of Merton.
      2nd Sire, Merton, Lord Walsing-
         ham's.

155 Ram, born in 1891.
      Sire, a Son of Merton.
      2nd Sire, Merton, Lord Walsing-
         ham's.

156 Ram, born in 1891.
      Sire, a Son of Merton.
      2nd Sire, Merton, Lord Walsing-
         ham's.

Flock
Book No.

157 Ram, born in 1891.
　　Sire, a Son of Merton.
　　2nd Sire, Merton, Lord Walsing-
　　　　ham's.

(*Continued on page* 51.)

———

Entered by Sir THOMAS BARRETT-LENNARD, Bart., Woodingdean, Brighton.

FLOCK No. 7.

158 Ram, born in 1889, "**Bartlow**,"
　　　　Webb's No. 6 of 1890.
　　1st Sire, Young Melton (Webb's).　　Dam, a Webb Ewe.　Sire,
　　2nd Sire, Melton (Webb's).　　　　　Webb's No. 22 in 1882.
　　3rd Sire, St. Blaize (Webb's).
　　4th Sire, Derby (Webb's).

159 Ram, born in 1889, "**Adventurer**,"
　　　　Webb's No. 25 of 1890.
　　1st Sire, Enterprise (Webb's).　　Dam, a Webb Ewe.　Sire,
　　2nd Sire, Gloucester (Webb's).　　 Webb's 53 in 1885.
　　3rd Sire, General Favourite (Webb's).
　　4th Sire, No. 20 in 1880 (Webb's).

160 Ram, born in 1889, "**Audley**" Webb's
　　　　No. 33 of 1890.
　　1st Sire, Young Melton (Webb's).　　Dam, a Webb Ewe.　Sire,
　　2nd Sire, Melton (Webb's).　　　　　Webb's 27 in 1885.
　　3rd Sire, St. Blaize (Webb's).
　　4th Sire, Derby (Webb's).

161 Ram, born in 1889, "**Cambridge**,"
　　　　Webb's No. 46 of 1890.
　　1st Sire, Cambridgeshire (Webb's). Dam, a Webb Ewe.　Sire,
　　2nd Sire, Gloucester (Webb's).　　　Gloucester.
　　3rd Sire, General Favourite (Webb's).
　　4th Sire, No. 20 in 1880 (Webb's).

162 A Ram, born in 1889, "**Linton**,"
　　　　Webb's No. 90 in 1890.
　　1st Sire, Cambridgeshire (Webb's). Dam, a Webb Ewe.　Sire,
　　2nd Sire, Gloucester (Webb's).　　　Gloucester.
　　3rd Sire, General Favourite (Webb's).
　　4th Sire, No. 20 in 1880 (Webb's).

163 Webb Ram.
　　Bred by Henry Webb and pur-
　　　chased at his Sale in 1889, by
　　　J. J. Colman, M.P.

164 Ram. No. 8 in Ellis' 1890 Sale.

165 Ram, born in 1890, No. 43 in Ellis'
        1891 Sale.
        1st Sire, Essex 2nd.   Dam, an Ellis Ewe.
        2nd Sire, Essex.
        3rd Sire, Colman Ram.

166 Ram, born in 1890, "**Lennard's
        Warnham.**" No. 23 Lucas'
        1891 Sale.
        1st Sire, Webb's 58 in 1889 Sale.
        2nd Sire, Webb's General Favourite.
        3rd Sire, Webb's No. 20 in 1880.

167 Ram, born in 1891, "**Belhus.**"
        Sire, No. 164 in Flock Book.   Dam, a Warnham Court Ewe.

Entered by WILLIAM TOOP, Aldingbourne, Chichester.

FLOCK No. 9.

168 Ram, born in 1889, "**Nelson.**"
        Sire, Toop's 15 in 1888.   Dam, a Toop Ewe.
        2nd Sire, Penfold's 20 in 1885.
        3rd Sire, General Favourite.
        4th Sire, Webb's No. 20 in 1880.

169 Ram, born in 1890.
        By No. 168 in Flock Book.   Dam, a Toop Ewe.

170 Ram, born in 1890.
        Sire, Toop's Newport.   Dam, a Toop Ewe.
        2nd Sire, a Botting Ram.
        3rd Sire, a Colman Ram.

171 Ram, born in 1890, "**Laddie.**"
        Sire, Toop's Newport.   Dam, a Toop Ewe.
        2nd Sire, a Botting Ram.
        3rd Sire, a Colman Ram.

172 Ram, born in 1890. By No. 258 in
        Flock Book.   Dam, a Toop Ewe.

173 Ram, born in 1890. By No. 258 in
        Flock Book.   Dam, a Toop Ewe.

174 Ram, born in 1890. By No. 258 in
        Flock Book.   Dam, a Toop Ewe.

175 Ram, born in 1890. By No. 258 in
        Flock Book.   Dam, a Toop Ewe.

176 Ram, born in 1890. By No. 258 in
        Flock Book.   Dam, a Toop Ewe.

Flock
Book No.

177   Ram, born in 1890.   By No. 258 in
      Flock Book.   Dam, a Toop Ewe.

178   Ram, born in 1890.   By No. 258 in
      Flock Book.   Dam, a Toop Ewe.

179   Ram, born in 1890.   By No. 258 in
      Flock Book.   Dam, a Toop Ewe.

180   Ram, born in 1890.
      Sire, No. 205 in Flock Book.   Dam, a Toop Ewe.

181   Ram, born in 1890.
      Sire, " **Tunbridge Wells.**"   Dam, a Toop Ewe.
      2nd Sire, No. 20, Penfold, 1888.
      3rd Sire, Webb Ram.

182   Ram, born in 1891.
      Sire, No. 171 in Flock Book.   Dam, a Toop Ewe.

183   Ram, born in 1891.
      Sire, No. 299 in Flock Book.   Dam, a Gibson Ewe.

184   Ram, born in 1891.
      Sire, No. 299 in Flock Book.   Dam, a Toop Ewe.

185   Ram, born in 1891.
      Sire, No. 299 in Flock Book.   Dam, a Toop Ewe.

186   Ram, born in 1891.
      Sire, Newman Ram, No. 185 at
        Chichester 1890 Sale.   Dam, a Toop Ewe.
      2nd Sire, Humphrey Ram.

187   Ram, born in 1891.
      Sire, Newman Ram, No. 185 at
        Chichester 1890 Sale.   Dam, a Toop Ewe.
      2nd Sire, Humphrey Ram.

188   Ram, born in 1891.
      Sire, Newman Ram, No. 185 at
        Chichester 1890 Sale.   Dam, a Toop Ewe.
      2nd Sire, Humphrey Ram.

189   Ram, born in 1891.
      Sire, Son of No. 258 in Flock
        Book.   Dam, a Toop Ewe.

190   Ram, born in 1891.
      Sire, Son of No. 258 in Flock
        Book.   Dam, Toop Ewe.

191   Ram, born in 1891.
      Sire, Son of No. 258 in Flock
        Book.   Dam, Toop Ewe.

Flock
Book No.

192   Ram, born in 1891.
        Sire, Son of No. 258 in Flock
          Book.   Dam, Toop Ewe.

193   Ram, born in 1891.
        Sire, No. 299 in Flock Book.   Dam, Toop Ewe.

194   Ram, born in 1891.
        Sire, No. 171 in Flock Book.   Dam, Toop Ewe.

195   Ram, born in 1891.
        Sire, No. 171 in Flock Book.   Dam, Toop Ewe.

196   Ram, born in 1891.
        Sire, Newman Ram, No. 185 at
          Chichester Sale, 1890.   Dam, Toop Ewe, by Webb
        2nd Sire, Humphry Ram.   Ram.

197   Ram Lamb, born in 1892.
        Sire, Son of No. 299 in Flock
          Book.   Dam, Toop Ewe.

198   Ram Lamb, born in 1892.
        Sire, No. 173 in Flock Book.   Dam, Toop Ewe.

199   Ram Lamb, born in 1892.
        Sire, No. 173 in Flock Book.   Dam, Toop Ewe.

200   Ram Lamb, born in 1892.
        Sire, No. 173 in Flock Book.   Dam, Toop Ewe.

201   Ram Lamb, born in 1892.
        Sire, No. 173 in Flock Book.   Dam, Toop Ewe.

202   Ram Lamb, born in 1892.
        Sire, No. 173 in Flock Book.   Dam, Toop Ewe.

203   Ram Lamb, born in 1892
        Sire, No. 173 in Flock Book.   Dam, Toop Ewe.

204   Ram Lamb, born in 1892.
        Sire, Grandson of No. 205 in
          Flock Book.   Dam, Toop Ewe.

205   Toop's famous No. 5.
        1st Sire, Penfold's No. 20 in 1885. Dam, Toop Ewe.
        2nd Sire, General Favourite.
        3rd Sire, Webb's No. 20 in 1880.

(*Continued on page* 51.)

Entered by H. F. LOCKE-KING, Brooklands, Weybridge.

FLOCK No. 10.

206 Ram, born in 1886 (1), Penfold 36
in 1888.
1st Sire, Young Crawford. Dam, Penfold Ewe.
2nd Sire, Crawford.
3rd Sire, Ram Lamb, No. 49
Lord Walsingham's 1st sale.

207 Ram, born in 1888 (2).
1st Sire, Colman's Old Champion. Dam, Colman's Ewe by Son of
2nd Sire, Grandson of Colman's Royal Kilburn.
Birmingham.

208 Ram, born in 1890, Ellis 35 in 1891 (3).
1st Sire, Earl. Dam, Ellis Ewe.
2nd Sire, Merton.
3rd Sire, Royal Reading.

209 Ram, Colman's No. 3.
Sire, Old Norwich. Dam, Colman Ewe by Webb's
Norfolk Favourite.

210 Ram, born in 1890 (5).
Sire, No. 209 in Flock Book. Dam, Ellis Ewe.

211 Ram, born in 1891 (6).
Sire, No. 206 in Flock Book. Dam, Onslow Ewe.

212 Ram, born in 1891 (7).
Sire, No. 207 in Flock Book. Dam, Onslow Ewe.

213 Ram, born in 1891 (8).
Sire, No. 207 in Flock Book. Dam, Onslow Ewe.

214 Ram, born in 1891 (11).
Sire, No. 206 in Flock Book. Dam, Oakhurst Ewe.

————

Entered by E. ELLIS, Shalford, Guildford.
(Numbers in brackets ( ) are ear numbers.)

FLOCK No. 12.

215 Ram, " Baronet "
1st Sire, Ripon. Dam, by a Son Dam, Throckmorton Ewe.
of Royal Taunton.
2nd Sire, Royal Reading.

44

216 Ram, born in 1890, "**Warwick.**"
    1st Sire, Webb's 35 in 1889. Dam   Dam, Ellis Ewe by Baronet.
      by Melton.
    2nd Sire, General Favourite. Dam
      by Flack's Favourite.
    3rd Sire, Webb's No. 20 in 1880.

217 Ram, born in 1890, "**Wye.**"
    1st Sire, Prince.   Dam. E'lis Ewe by Colman's
    2nd Sire, No. 215 in Flock Book.   No. 18.

218 Ram, born in 1891, "**Arnold**" (552).
    1st Sire, No. 215 in Flock Book.   Dam (360) Ellis Ewe.

219 Ram, born in 1891, "**Aden**" (468).
    1st Sire, Colman's 18.   Dam (265) Ellis Ewe by
    2nd Sire, Webb's No. 47.   Throckmorton Ram.

220 Ram, born in 1891, "**Artist**" (469).
    1st Sire, Webb's 35 in 1889. Dam   Dam, Lucas Ewe.
      by Melton.
    2nd Sire, General Favourite. Dam
      by Flack's Favourite.
    3rd Sire, Webb's No. 20 in 1880.

221 Ram, born in 1891, "**Author**" (377).
    1st Sire, Royal Plymouth.   Dam, Ellis Ewe (102) by
    2nd Sire, No. 215 in Flock Book.   Botting Ram.

222 Ram, born in 1891, "**Archie**" (359).
    Sire, Duke.   Dam (259), Ellis Ewe.
    2nd Sire, Merton.
    3rd Sire, H. C. Ram (York).
    4th Sire, Royal Reading.

223 Ram, born in 1891, "**Allen**" (437).
    Sire, Webb's 35 in 1889. Dam by
      Melton.   Dam (55), Botting Ewe.
    2nd Sire, General Favourite.
    3rd Sire, Webb's No. 20 in 1880.

224 Ram, born in 1891, "**Asia**" (533).
    Sire, No. 215 in Flock Book.   Dam (2), Ellis Ewe, by Throck-
      morton Ram.

225 Ram, born in 1891, "**Azure**" (582).
    Sire, No. 215 in Flock Book.   Dam (54), Botting Ewe.

226 Ram, born in 1891, "**Arab**" (152).
    Sire, Kingson.   Dam (255), Ellis Ewe, by
    2nd Sire, King.   Ripon.
    3rd Sire, Throckmorton Ram.
      Dam, Walsingham Ewe.

Flock
Book No.

227 Ram, born in 1891, " **Ant** " (561).
    Sire, No. 215 in Flock Book.   Dam (128), Ellis Ewe.

228 Ram, born in 1891, " **Adair** " (470).
    Sire, Son of Newcastle.   Dam, Lucas Ewe.
    2nd Sire, Royal Newcastle.
    3rd Sire, Merton.

229 Ram, born in 1891, " **Arc** " (577).
    Sire, Baron.   Dam (335), Ellis Ewe, by
    2nd Sire, No. 215 in Flock Book.   Merton.

230 Ram, born in 1891, " **Abel** " (419).
    Sire, Webb's 35 in 1889.   Dam, Lucas Ewe.
    2nd Sire, General Favourite.
    3rd Sire, Webb's No. 20 in 1880.

231 Ram, born in 1891, " **Adieu** " (412).
    Sire, Webb's 35 in 1889.   Dam, Brassey Ewe.
    2nd Sire, General Favourite.
    3rd Sire, Webb's No. 20 in 1880.

232 Ram, born in 1891, " **Abbot** " (578).
    Sire, Son of Duke.   Dam, Lucas Ewe.
    2nd Sire, Duke.
    3rd Sire, Son of Merton.

233 Ram, born in 1891, " **Aln** " (509).
    Sire, Son of Duke.   Dam (257), Ellis Ewe.
    2nd Sire, Duke.
    3rd Sire, Merton.

234 Ram, born in 1891, " **Alnwick** " (522).
    Sire, Baron.   Dam (262), Ellis Ewe by
    2nd Sire, No. 215 in Flock Book.   Throckmorton Ram.

235 Ram, born in 1891, " **Aloft** " (366).
    Sire, Royal Plymouth.   Dam (159), Ellis Ewe by
    2nd Sire, No. 215 in Flock Book.   Botting Ram.

236 Ram, born in 1891, " **Amos** " (507).
    Sire, Royal Newcastle.   Dam (81), Ellis Ewe.
    2nd Sire, Merton.

237 Ram, born in 1891, " **Avon** " (472).
    Sire, No. 215 in Flock Book.   Dam, Brassey Ewe.

(*Continued on page 56.*)

Entered by W. FLUX, Esq., Ampney Crucis, Cirencester.

FLOCK No. 14.

238 Ram, born in 1891.
    Sire, No. 57 in Flock Book.   Dam, an Ewe, bred by Earl Bathurst.

239 Ram, born in 1891.
    Sire, No. 57 in Flock Book.   Dam, an Ewe, bred by Earl Bathurst.

240 Ram, born in 1891.
    Sire, No. 57 in Flock Book.   Dam, an Ewe, bred by Earl Bathurst.

241 Ram, born in 1891.
    Sire, No. 57 in Flock Book.   Dam, an Ewe, bred by Earl Bathurst.

242 Ram, born in 1891.
    Sire, No. 57 in Flock Book.   Dam, an Ewe, bred by Earl Bathurst.

243 Ram, born in 1891.
    Sire, No. 57 in Flock Book.   Dam, an Ewe, bred by Earl Bathurst.

244 Ram, born in 1891.
    Sire, No. 57 in Flock Book.   Dam, an Ewe, bred by Earl Bathurst.

Entered by H. F. LOCKE-KING, Brooklands, Weybridge.

FLOCK No. 11.

245 Ram, born in 1889 (27 of Webb's 1890) (1).
    Sire, Cambridgeshire.   Dam, Webb Ewe, by Derby.
    2nd Sire, Gloucester.
    3rd Sire, General Favourite.
    4th Sire, Webb's No. 20 in 1880.

246 Ram, born in 1890, "**Weybridge**" No. 2.
    1st Sire, Webb's No. 24 of 1889.
    Dam by Hardihood.   Dam, Lythe Hill Ewe.
    2nd Sire, Little Twin.   Dam
      by St. Blaize.
    3rd Sire, General Favourite.   Dam
      by Flack's Favourite.
    4th Sire, No. 20 in 1880.

247 Ram, born in 1891, "**Weybridge**" No. 3.
Sire, Favourite.

248 Ram, born in 1890, "**Weybridge**" No. 4.
Bred by H. F. Locke-King at
Oakhurst.

———

Entered by EARL BATHURST, Cirencester.

(*Numbers in brackets ( ) are ear numbers.*)

FLOCK No. 15.

249 "**Old Twenty**" (720).
    1st Sire, Colman Ram.   Dam, Colman Ewe.
    2nd Sire, Colman's Kilburn.

250 "**Witley**" (721).
    Sire, No. 249 in Flock Book.   Dam, Bathurst Ewe.

251 "**Johnnie II.**" (725).
    1st Sire, Johnnie.   Dam, Bathurst Ewe.
    2nd Sire, Colman Ram.
    3rd Sire, Webb's Norfolk Favourite.

252 Ram (726).
    Sire, No. 251 in Flock Book.   Dam, Bathurst Ewe.

253 Ram (727)
    Sire, No. 251 in Flock Book.   Dam, Bathurst Ewe.

254 Ram, born in 1888, Webb's 34 in 1889.
    Sire, Webb's General Favourite.   Dam, Webb Ewe.
    2nd Sire, Webb's 20 in 1880.

255 Ram, born in 1888, Webb's 66 in 1889.
    Sire, Webb's Little Twin.   Dam, Webb Ewe, by Hardihood.
    2nd Sire, Webb's General Favourite.

256 Ram (703).

257 Ram (715).

258 Ram (Toop's famous No. 8).
    Sire, Son of Penfold's No. 20.   Dam, Toop Ewe.
    2nd Sire, Penfold's No. 20 in 1885.
    3rd Sire, Webb's General Favourite.
    4th Sire, Webb's No. 20 in 1880.

259 Ram (Toop's No. 10 in 1890).
    Sire, Newport.   Dam, Toop Ewe.
    2nd Sire, Botting Ram.
    3rd Sire, Colman Ram.

Entered by HARRY HABIN, Birdham, Chichester.

FLOCK No. 5.

260 "Shortlegs," born in 1886, No. 7 in Webb's 1889 Sale.
Sire, Foljambe.   Dam, Webb Ewe.   By No. 41
2nd Sire, St. Blaize.   in 1879.
3rd Sire, Derby.
4th Sire, Shepherd's Favourite.

261 Ram, bred by F. N. Hobgen, born in 1887.
Sire, Ram bred by F. N. Hobgen.   Dam, Hobgen Ewe.

262 Ram, bred by H. Penfold, born in 1889.
Sire, Duke of Richmond's No. 10
     in 1888.   Dam, Penfold Ewe.

263 Ram bred by the Duke of Richmond, born in 1890.
Sire, Cambridgeshire.   Dam, Goodwood Ewe.
2nd Sire, Gloucester.
3rd Sire, General Favourite.
4th Sire, Webb's No. 20 in 1880.

264 Ram, born in 1891, by Hobgen Ram.   Dam, Habin Ewe.

265 Ram, born in 1891, by Hobgen Ram.   Dam, Habin Ewe.

Entered by JAMES BLYTH, Blythwood, Stanstead, Essex.

(Numbers in brackets ( ) are ear numbers.)

FLOCK No. 18.

266 Ram, born in 1886, "**General.**" Lot 4, Webb's 1889 Sale.
Sire, General Favourite.   Dam by   Dam, Webb Ewe.   Sire, No.
    Flack's Favourite.   23 in 1883.
2nd Sire, No. 20 in 1880.

267 Ram, born in 1887, "**King of the Castle.**" Lot 15, Webb's 1889
    Sale.
Sire, Delorme.   Dam by No. 20   Dam, Webb Ewe.   Sire, No.
    in 1883.   28 in 1883.
2nd Sire, No. 27 in 1884.

268 Ram, born in 1888, "**Young Gloucester**" (312). Lot 32, Webb's
    1889 Sale.
Sire, No. 51 in Flock Book.   Dam, Webb Ewe.   Sire, Shep-
                             herd's Favourite.

269 Ram, born in 1889, **"Streetly Hall Favourite"** (38). Lot 2, Webb's
     1890 Sale.
     Sire, Cambridgeshire. Dam by Dam, Webb Ewe. Sire,
     St. Blaize.   General Favourite.
     2nd Sire, No. 51 in Flock Book.

270 Ram, born in 1889, **"Double Gloucester"** (13). Lot 32, Webb's
     1890 Sale.
     Sire, No. 51 in Flock Book.   Dam, Webb Ewe. Sire,
                                     Derby.

271 Ram, born in 1890 (158).
     Sire, Colman Ram.   Dam, Colman Ewe. Sire,
                                     Webb's 41 in 1886.

272 Ram, born in 1890 (139).
     Sire, Plymouth Champion (Col- Dam, Colman Ewe. Sire,
     man).   A Son of Norfolk
     2nd Sire, Colman Ram.   Favourite.

273 Ram, born in 1891, **"Aldingbourne"** (51).
     Sire, Colman's Plymouth Cham- Dam, 341. Sire, Webb's
     pion.   Melton.
     2nd Sire, Colman Ram.

274 Ram, born in 1889, **"Sir William"** (oo).
     Sire, Nottingham. Dam, Good- Dam, a Throckmorton Ewe.
     wood Ewe. (See No. 57 in
     Flock Book.)

275 Ram, born in 1890 (34).
     Sire, No. 267 in Flock Book.   Dam, Webb Ewe (143). Sire,
                                     Webb's Hardihood.

276 Ram (59), born in 1891.
     Sire, No. 267 in Flock Book.   Dam, Webb Ewe (305). Sire,
                                     Webb's Hardihood.

277 Ram (22), born in 1890.
     Sire, No. 268 in Flock Book.   Dam, Webb Ewe (341). Sire,
                                     Webb's Melton.

278 Ram (27), born in 1890.
     Sire, No. 268 in Flock Book.   Dam, Webb Ewe (338).
                                     Sire, Webb's Melton.

279 Ram (56), born in 1891.
     Sire, No. 269 in Flock Book.   Dam, Ewe (167). Sire, J. J.
                                     Colman's Ram, "Penfold."

Entered by Mrs. PERKINS, Saham Hall, Wotton, Norfolk.

(Numbers in brackets ( ) are ear numbers.)

FLOCK No. 19.

Flock
Book No.

280 Ram, born 1887, "**Banker**" (18), bred
by H. Webb.
1st Sire, General Favourite. Dam,   Dam, Webb Ewe. Sire, Derby.
by Flack's Favourite.
2nd Sire, No. 20 in 1880.

281 Ram, born 1888, "**Little Peru**" (266),
bred by H. Webb.
1st Sire, Golding's Dam by   Dam, Webb Ewe. Sire,
Shepherd's Favourite.   Guinea Peru.
2nd Sire, Hardihood. Dam by
Throckmorton Favourite.
3rd Sire, Baltimore 1st.

282 Ram, born 1889, "**Ickworth**" (246),
bred by H. Webb.
1st Sire, Webb's Ickworth Favourite. Dam, a Webb Ewe. Sire,
Dam by Derby.   Norfolk Favourite.
2nd Sire, H. C. Webb's Favourite.

283 Ram, born 1889, "**Perkins Badmington**"
(333), bred by H. Webb.
Sire, Webb's Badmington. Dam,   Dam, Webb Ewe. Sire, No.
by St. Blaize.   27 in 1885.
2nd Sire, Gloucester, 51, Flock Book.

284 Ram, "**Credit**" (1), born in 1891.
Sire, No. 281, Flock Book.   Dam, a Merton Ewe.

———

**Entered by Messrs. DE MURIETTA, Wadhurst Park,
Wadhurst, Sussex.**

(Continued from page 31.)

FLOCK No. 1.

290 Ram (6 tooth), W.P. 44 of 1889.
Sire, No. 48 in Flock Book.   Dam (95 left), Penfold Ewe, by
Penfold's Brighton Champion.

Entered by PAGHAM HARBOUR COMPANY, Selsey, Chichester.

(*Continued from page* 34.)

FLOCK No. 3.

Flock
Book No.

291 "**Maidstone**," born in 1884.
Sire, Penfold's 39 in 1883.　　Dam, Pagham Harbour Ewe.
2nd Sire, No. 23 in 1875.
3rd Sire, No. 343 in Flock Book.

292 "**Selsey**" (292), bred by Penfold, born 1885.
Sire, Webb's General Favourite.　　Dam, Penfold Ewe.
2nd Sire, Webb's No. 20 of 1880.

293 "**Beacon 3rd**" (293), born in 1890.
Sire, No. 71 in Flock Book.　　Dam, Pagham Harbour Ewe.

294 "**Padwick**" (294), born in 1891, bred by H. Padwick.
Sire, No. 330 in Flock Book.　　Dam, Thorney Ewe, by 314.
　　　　　　　　　　　　　　　　　Dam by 308.

295 "**Heasman**" (295), born in 1891.
Sire, No. 62 in Flock Book.　　Dam, Heasman Ewe.

296 "**Royal Nottingham 5th**" (296), born in 1891.
Sire, No. 62 in Flock Book.　　Dam, Pagham Harbour Ewe.

297 "**Champion 2nd**," bred by Penfold.　121 at sale.　Born in 1888.
Sire, Penfold Ram.　　　　　　　　Dam, Penfold Ewe.
2nd Sire, Penfold's Brighton Cham-
　　　pion, 1885.
3rd Sire, Penfold Ram.
4th Sire, Ram bred by Captain
　　　Taylor, by No. 342 in Flock
　　　Book.

Entered by F. N. HOBGEN, Appledram.

(*Continued from page* 39.)

FLOCK No. 6.

298 A Ram, born in 1890.
Sire, No. 266 in Flock Book.　　Dam, Hammond Ewe.

Entered by W. TOOP, Aldingbourne, Chichester.

(*Continued from page* 42.)

FLOCK No. 9.

299 "**Little Hero**."
Sire, Newport.　　　　　　Dam, a Toop Ewe.
2nd Sire, Botting Ram.
3rd Sire, Colman Ram.

300   Ram Lamb, born in 1892.
      Sire, No. 184 in Flock Book.   Dam, a Toop Ewe.

301   Ram Lamb, born in 1892.
      Sire, No. 304 in Flock Book.   Dam, a Toop Ewe.

302   Ram Lamb, born in 1892.
      Sire, No. 304 in Flock Book.   Dam, a Toop Ewe.

303   Ram Lamb, born in 1892.
      Sire, No. 304 in Flock Book.   Dam, a Toop Ewe.

304   Ram, "**Waterbeach.**"
      Sire, Grandson of Duke of
         York.   Dam, a Goodwood Ewe.
      3rd Sire, Duke of York.
      4th Sire, Buckland.

Entered by H. PADWICK, Thorney, Emsworth.

FLOCK No. 16.

305   Ram, bred by H. P. Hart.
      Sire, 337 in Flock Book.   Dam, a Ewe bred by H. P. Hart.
                           1st Sire, a Goodwood Ram
                             (No. 12). Dam, Hart Ewe.
                           2nd Sire, Goodwood Ram
                           (No. 10).

306   Ram, born in 1878, bred by H. P. Hart.
      1st Sire, a Hart Ram.   Dam, Hart   Dam, a Ewe bred by H. P. Hart.
      Ewe.   1st Sire, Goodwood Ram
      2nd Sire, No. 337 in Flock Book.   (No. 12). Dam, Hart Ewe.
                           2nd Sire, Goodwood Ram
                           (No. 10).

307   Ram, born in 1878, bred by H. Penfold.
      Sire, Webb's Marquis.   Dam, a Penfold Ewe.
                         Sire, a Ram bred by H. P. Hart.

308   Ram, born in 1878, Thorney No. 3.
      Sire, No. 305 in Flock Book.   Dam, a Thorney Ewe.

309   Ram, born in 1881, Thorney No. 8.
      Sire, No. 306 in Flock Book.   Dam, a Hart Ewe.

310   Ram, born in 1878, bred by H. Penfold.
      Sire, Ram Lamb No. 49, Lord   Dam, a Penfold Ewe.
      Walsingham at his first sale.

311   Ram, born in 1880, Thorney No. 1.
      Sire, No. 310 in Flock Book.   Dam, a Hart Ewe.

312 Ram, born in 1880, Thorney No. 2.
Sire, No. 310 in Flock Book.   Dam, a Hart Ewe.

313 Ram, born in 1884, Thorney No. 5.
Sire, No. 309 in Flock Book.   Dam, a Thorney Ewe.
Sire, No. 305 in Flock Book.

314 Ram, born in 1884, Thorney No. 6.
Sire, No. 309 in Flock Book.   Dam, a Thorney Ewe.
Sire, No. 305 in Flock Book.

315 Ram, born in 1884, Thorney No. 7.
Sire, No. 309 in Flock Book.   Dam, a Thorney Ewe.
Sire, No. 305 in Flock Book.

316 Ram, bred by Rigden.   Dam, a Rigden Ewe.

317 Ram, bred by Duke of Richmond.
Sire, Goodwood Ram.
2nd Sire, a Rigden Ram.   Dam, a Goodwood Ewe.

318 Ram, born in 1883, Thorney No. 10.
Sire, No. 311 in Flock Book.   Dam, a Thorney Ewe.
Sire, No. 309 in Flock Book.

319 Ram, born in 1885, Thorney No. 13.
Sire, No. 316 in Flock Book.   Dam, a Thorney Ewe.
1st Sire, No. 308 in Flock Book.
2nd Sire, No. 307 in Flock Book.

320 Ram, born in 1881, Thorney No. 11.
Sire, No. 307 in Flock Book.   Dam, a Thorney Ewe.
Sire, No. 305 in Flock Book.

321 Ram, born in 1881, Thorney No. 12.
Sire, No. 307 in Flock Book.   Dam, a Thorney Ewe.
Sire, No. 305 in Flock Book.

322 Ram, born in 1885, Thorney No. 16.
Sire, No. 308 in Flock Book.   Dam, a Thorney Ewe.
Sire, No. 306 in Flock Book.

323 Ram, born in 1885, Thorney No. 17.
Sire, No. 308 in Flock Book.   Dam, a Thorney Ewe.
Sire, No. 306 in Flock Book.

Flock
Book No.

324 Ram, born in 1887, Thorney No. 101.
      Sire, No. 318 in Flock Book.   Dam, a Thorney Ewe.
                                    Sire, No. 309 in Flock Book.
                                        Dam, by No. 320 in Flock Book.

325 Ram, born in 1887, Thorney No. 103.
      Sire, No. 318 in Flock Book.   Dam, a Thorney Eve.
                                      Sire, No. 309 in Flock Book.
                                        Dam, by No. 320 in Flock Book.

326 Ram, born in 1885, Thorney No. 110.
      Sire, No. 320 in Flock Book.   Dam, a Thorney Ewe.
                                      Sire, No. 306 in Flock Book.

327 Ram No 84, Penfold's 1888 Sale.
      Sire, Penfold's Brighton Champion,
         1885.   Dam, a Penfold Ewe.

328 Ram bred by Robert Drewitt.
      1st Sire, Emery Ram.   Dam, a Drewitt Ewe.
      2nd Sire, Rigden Ram, out of
         Emery Ewe.   1st Sire, Humphrey Ram.
                                        Dam, Hart Ewe.
                                      2nd Sire, No. 338 in Flock Book.

329 Ram, bred by Viscount Gage (sold at his 1888 Sale).
      Sire, a Penfold Ram.   Dam, a Gage Ewe.
                                      Sire, Ram No. 16, Hart's Sale, 1878.
                                      Dam by a Walderton Ram, bred by J. Pinnix.

330 Ram, born in 1886, No. 1 Viscount Gage's Sale, 1888.
      Sire, a Penfold Ram.   Dam, a Hampden Ewe.
                                      Sire, Emery Ram.

331 Ram, bred by H. P. Hart.
      Sire, No. 337 in Flock Book.   Dam, a Hart Ewe.

332 Ram, bred by Mr. Rigden.
      Sire, No. 333 in Flock Book.

333 Ram, No. 63 Jonas Webb's Sale, 1862.
      1st Sire, Lot 1 Jonas Webb's 1861
         Sale.   Dam, a Jonas Webb Ewe.
      2nd Sire, Young Elegance (No.   1st Sire, No. 23 in 1861.
         246 of 1856). Dam by Jonas   2nd Sire, The Frenchman.
         Webb's Old 70.   3rd Sire, Jonas Webb's 2nd prize Shearling at Gloucester.

334 Ram.
      Sire, No. 309 in Flock Book.   Dam, a Madgwick Ewe.
                                        Sire, No. 321 in Flock Book.

Flock
Book No.

335 Ram, born in 1891.
Sire, No. 330 in Flock Book.   Dam, a Thorney Ewe.
Sire, No. 314 in Flock Book.
Dam by No. 308 in Flock
Book.

336 Ram, born in 1891.
Sire, No. 330 in Flock Book.   Dam, a Thorney Ewe.
Sire, No. 314 in Flock Book.
Dam by No. 308 in Flock
Book.

337 Ram, bred by Rigden. 1st prize R.S.A.E., Hull, 1872.
Sire, No. 333 in Flock Book.   Dam, a Rigden Ewe.

338 Ram, bred by H. P. Hart.
Dam, a Hart Ewe.

339 Ram, born in 1888.
Sire, No. 309 in Flock Book.   Dam, a Thorney Ewe.
Sire, No. 318 in Flock Book.
Dam by No. 308 in Flock
Book.

340 Ram, bred by Mrs. Hart.
Sire, No. 309 in Flock Book.   Dam, Hart Ewe by 321 in
Flock Book.

————

Entered by HUGH PENFOLD, Selsey, Chichester.

(*Continued from page* 37.)

FLOCK No. 4.

341 " Brighton Champion."
1st Sire, Penfold Ram.   Dam, Woodman Ewe by Penfold
2nd Sire, Ram, bred by Captain   Ram.
Taylor.
3rd Sire, No. 342 in Flock Book.

342 Lord Walsingham's No. 25 in 1871.
1st Sire, Royal Manchester, Dam   Dam, Merton Ewe.
by Hart Ram.   Sire, Webb's 107.
2nd Sire, Young Gaiety, Dam by   2nd Sire, Salisbury.
Ellman Ram.
3rd Sire, Webb's No. 102.
4th Sire, Webb's Archbishop,
Dam by Second York.

343 Ram Lamb, born in 1871. No. 49 at Lord Walsingham's sale, 1871.
   1st Sire, No. 19 in 1871 sale.   Dam, Merton Ewe.
   2nd Sire, Royal Worcester, Dam   1st Sire, Son of Clipper.
     by Hart Ram.   2nd Sire, Clipper, Dam by
     Ellman.
     3rd Sire, Webb's 102.
     4th Sire, Webb's Archbishop.

344 Ram, 39 in 1883.
   1st Sire, No. 23 in 1875.   Dam, Penfold Ewe.
   2nd Sire, No. 343 in Flock Book.

Entered by EDWIN ELLIS, Shalford, Guildford.

(*Continued from page* 45)

FLOCK No. 12.

345 Ram, " **Ripon.** "
   Sire, Royal Reading.   Dam, Merton Ewe.
    Dam by Son of Royal Taunton.

346 Ram, " **Merton.** "
   1st Sire, H. C. Shearling York.   Dam, Merton Ewe.
   2nd Sire, Royal Reading.
    Dam by Grandson of Royal Taunton.

347 Ram, " **Royal Newcastle.** "
   Sire, No. 346 in Flock Book.   Dam, Ellis Ewe.

348 Ram, " **Royal Plymouth.** "
   Sire, No. 215 in Flock Book.   Dam, Merton Ewe.

349 Ram, Webb's 35 in 1889 (344).
   1st Sire, General Favourite.   Dam, Webb Ewe by Melton.
    Dam by Flack's Favourite.
   2nd Sire, No. 20 in 1880.

EWES AND EWE LAMBS.

Entered by Messrs. DE MURIETTA, Wadhurst Park, Wadhurst, Sussex.

FLOCK No. 1.

(N.B.—Numbers in brackets ( ) are ear numbers. All numbers are in right ear unless otherwise stated.)

EWE PEDIGREES.

Flock
Book No.

1 Ewe (351) born 1891.
Sire is No. 52 in Flock Book.

Dam(424 left) Colman Ewe.
  1st Sire, Colman's Old Norwich.
    Dam, Colman Ewe.
  2nd Sire, Webb's Derby.
  3rd Sire, Webb's Shepherd's Favourite.

2 Ewe (380) born 1891.
Sire is No. 53 in Flock Book.

Dam (318 left) Webb Ewe.
  1st Sire, Webb's Ormonde.
    Dam by Hardihood.
  2nd Sire, Webb's General Favourite.
  3rd Sire, Webb's No. 20 in 1880.

3 Ewe (388) born 1891.
Sire is No. 55 in Flock Book.

Dam (502 left) Horsted Ewe.
  1st Sire, Webb's 26 in 1885.
    Dam, Horsted Ewe.
  2nd Sire, Webb's St. Blaize.
  3rd Sire, Webb's Derby.
  4th Sire, Webb's Shepherd's Favourite.

Entered by Mrs. PERKINS, Saham Hall, Wotton, Norfolk.

FLOCK No. 19.

4 Ewe (1).
Sire, a Merton Ram, bred by Lord Walsingham.   Dam, a Merton Ewe, bred by Lord Walsingham.

58

5 Ewe (2).
Sire, a Merton Ram, bred by Lord Walsingham.
Dam, a Merton Ewe, bred by Lord Walsingham.

6 Ewe (3).
Sire, a Merton Ram, bred by Lord Walsingham.
Dam, a Merton Ewe, bred by Lord Walsingham.

7 Ewe (4).
Sire, a Merton Ram, bred by Lord Walsingham.
Dam, a Merton Ewe, bred by Lord Walsingham.

8 Ewe (5), born in 1890.
Sire, Banker, No. 281 in Flock Book.
Dam, an Ewe by a Merton Ram. (Merton Ram bred by Lord Walsingham.)

9 Ewe (6), born in 1890.
Sire, Banker, No. 281 in Flock Book.
Dam, an Ewe by a Merton Ram. (Merton Ram bred by Lord Walsingham.)

10 Ewe (7), born in 1890.
Sire, Banker, No. 281 in Flock Book.
Dam, an Ewe by a Merton Ram. (Merton Ram bred by Lord Walsingham.)

11 Ewe (8), born in 1890.
Sire, Banker, No. 281 in Flock Book.
Dam, an Ewe by a Merton Ram. (Merton Ram bred by Lord Walsingham.)

12 Ewe (9), born in 1890.
Sire, Banker, No. 281 in Flock Book.
Dam, an Ewe by a Merton Ram. (Merton Ram was bred by Lord Walsingham.)

13 Ewe (10), born in 1890.
Sire, Banker, No. 281 in Flock Book.
Dam, an Ewe by a Merton Ram. (Merton Ram was bred by Lord Walsingham.)

14 Ewe (11), born in 1890.
Sire, Banker, No. 281 in Flock Book.
Dam, an Ewe by a Merton Ram. (Merton Ram was bred by Lord Walsingham.)

15 Ewe (12), born in 1890.
Sire, Banker, No. 281 in Flock Book.
Dam, an Ewe by a Merton Ram. (Merton Ram was bred by Lord Walsingham.)

16 Ewe (13), born in 1890.
     Sire, Banker, No. 281 in Flock   Dam, an Ewe by a Merton Ram.
     Book.   (Merton Ram was bred by
                         Lord Walsingham.)

17 Ewe (14), born in 1890.
     Sire, Banker, No. 281 in Flock   Dam, an Ewe by a Merton Ram.
     Book.   (Merton Ram was bred by
                         Lord Walsingham.)

18 Ewe (15), born in 1890.
     Sire, Banker, No. 281 in Flock   Dam, an Ewe by a Merton Ram.
     Book.   (Merton Ram was bred by
                         Lord Walsingham.)

19 Ewe (16), born in 1891.
     Sire, Banker, No. 281 in Flock   Dam, an Ewe by a Merton Ram.
     Book.   (Merton Ram was bred by
                         Lord Walsingham.)

20 Ewe (17), born in 1891.
     Sire Banker, No. 281 in Flock
     Book.   Dam, an Ewe by a Merton
                      Ram. (Merton Ram was
                      bred by Lord Walsingham.)

21 Ewe (18), born in 1891.
     Sire, Banker, No. 281 in Flock
     Book.   Dam, an Ewe by a Merton Ram.
                      (Merton Ram was bred by
                      Lord Walsingham.)

22 Ewe (19), born in 1891.
     Sire, Banker, No. 281 in Flock
     Book.   Dam, an Ewe by a Merton Ram.
                      (Merton Ram was bred by
                      Lord Walsingham.)

23 Ewe (20), born in 1891.
     Sire, Banker, No. 281 in Flock
     Book.   Dam, an Ewe by a Merton Ram.
                      (Merton Ram was bred by
                      Lord Walsingham.)

24 Ewe (21), born in 1891.
     Sire, Banker, No. 281 in Flock
     Book.   Dam, an Ewe by a Merton Ram.
                      (Merton Ram was bred by
                      Lord Walsingham.)

Flock
Book No.

25   Ewe (22), born in 1891.
      Sire, Banker, No. 281 in Flock
        Book.   Dam an Ewe by a Merton Ram.
                                  (Merton Ram was bred by
                                  Lord Walsingham.)

26   Ewe (23), born in 1891.
      Sire, Banker, No. 281 in Flock
        Book.   Dam, an Ewe by a Merton Ram.
                                  (Merton Ram was bred by
                                  Lord Walsingham.)

27   Ewe (24), born in 1891.
      Sire, Banker, No. 281 in Flock
        Book.   Dam, an Ewe by a Merton Ram.
                                  (Merton Ram was bred by
                                  Lord Walsingham.)

28   Ewe (25), born in 1891.
      Sire, Banker No. 281 in Flock
        Book.   Dam, an Ewe by a Merton Ram.
                                  (Merton Ram was bred by
                                  Lord Walsingham.)

29   Ewe (26), born in 1891.
      Sire, Banker No. 281 in Flock
        Book.   Dam, an Ewe by a Merton Ram.
                                  (Merton Ram was bred by
                                  Lord Walsingham.)

30   Ewe (27), born in 1891.
      Sire, Banker, No. 281 in Flock
        Book.   Dam, an Ewe by a Merton Ram.
                                  (Merton Ram was bred by
                                  Lord Walsingham.)

31   Ewe (28), born in 1891.
      Sire, Banker, No. 281 in Flock
        Book.   Dam, an Ewe by a Merton Ram.
                                  (Merton Ram was bred by
                                  Lord Walsingham.)

32   Ewe (29), born in 1891.
      Sire, Banker, No. 281 in Flock
        Book.   Dam, an Ewe by a Merton Ram.
                                  (Merton Ram was bred by
                                  Lord Walsingham.)

33   Ewe (30), born in 1891.
      Sire, Banker, No. 281 in Flock
        Book.   Dam, an Ewe by a Merton Ram.
                                  (Merton Ram was bred by
                                  Lord Walsingham.)

EWE LAMBS.

Entered by W. TOOP, Aldingbourne, Chichester.

FLOCK No. 9.

Flock
Book No.

34 Ewe Lamb, born in 1892.
  Sire, Son of No. 299.   Dam, an Ewe bred by W.
  2nd Sire, No. 299 in Flock Book. Toop.

35 Ewe Lamb, born in 1892.
  Sire, Son of No. 299.   Dam, an Ewe bred by W.
  2nd Sire, No. 299 in Flock Book. Toop.

36 Ewe Lamb, born in 1892.
  Sire, Grandson of No. 205 in Dam, an Ewe bred by W.
   Flock Book.   Toop.

37 Ewe Lamb, born in 1892.
  Sire, Grandson of No. 205 in Dam, an Ewe bred by W.
   Flock Book.   Toop.

38 Ewe Lamb, born in 1892.
  Sire, No. 172 in Flock Book. Dam, an Ewe bred by W.
            Toop.

Schedule of Prize Winners

(SOUTHDOWN),

AT THE

ROYAL AGRICULTURAL SHOW, WARWICK, 1892.

Two Shear Ram—

First Prize ...	...	...	DUKE OF RICHMOND.
Second Prize	...	...	EDWIN ELLIS.
Reserve	...	...	H.R.H. THE PRINCE OF WALES.

Shearing Ram—

First Prize ...	...	...	J. J. COLMAN.
Second Prize	...	...	DUKE OF HAMILTON.
Third Prize	...	...	A. DE MURIETTA.
Reserve	...	...	EDWIN ELLIS.

Pen of Three Ram Lambs—

First Prize ...	...	...	WILLIAM TOOP.
Second Prize	...	...	PAGHAM HARBOUR COMPANY.
Reserve	...	...	A. DE MURIETTA.

Pen of Three Shearling Ewes—

First Prize ...	...	...	J. J. COLMAN.
Second Prize	...	...	A. DE MURIETTA.
Third Prize	...	...	JAMES BLYTH.
Reserve	...	...	EDWIN ELLIS.

INDEX.

———•———

www.ingramcontent.com/pod-product-compliance
Lightning Source LLC
Chambersburg PA
CBHW081748220526
45468CB00008B/2285